ELGIN

TRENCH WATCHES OF THE GREAT WAR

STAN CZUBERNAT

Schiffer Publishing Ltd®
4880 Lower Valley Road • Atglen, PA 19310

Other Schiffer Books on Related Subjects:
Wristwatches: ***A Handbook and Price Guide, Revised 6th Edition,***
978-0-7643-2252-5, $19.95
American Wristwatches: ***Five Decades of Style and Design.***
978-0-7643-0171-1, $79.95
Breitling: ***The History of a Great Brand of Watches, 1884 to the Present.***
978-0-7643-2670-7, $49.95

Library of Congress Control Number: 2014954065

Designed by John P. Cheek
Cover Design by Justin Watkinson
Type set in House Movements/ITC New Baskerville Std

ISBN: 978-0-7643-4711-5
Printed in China

Published by Schiffer Publishing, Ltd.
4880 Lower Valley Road
Atglen, PA 19310
Phone: (610) 593-1777; Fax: (610) 593-2002
E-mail: Info@schifferbooks.com

I would like to dedicate this book to my beautiful wife, Tiffany. Her support and patience during this long project has made it all possible.

I would also like to thank the Elgin Area Historical Society in Elgin, Illinois, specifically Elizabeth Marston and Bill Briska, as well as The National Watch & Clock Museum in Columbia, Pennsylvania.

Other people whom I would like to thank include Richard Czubernat, Greg Ochs, Alvin Yin, Greg Hart, Fred Friedberg, Stephen Kowalski, Jessie Hueg, Eric Walton, and Adam Harris. Your contributions were greatly appreciated!

SELINGER'S
820-F.
SALE
ARMY WRIST
WATCHES
$4
A Good Watch
GENUINE LEATHER
KITCHENER
MILITARY
STRAP
10¢
20 J. SELINGER. 820

CONTENTS

CHAPTER 1

ABOUT ELGIN TRENCH WATCHES

You will find very few names next to the Elgin wristwatches in this book, simply because so few Elgin watches in this time period had proper names given to them. The Pershing, the Foch, and the Kitchener were the very first "named" Elgin men's wristwatches from the factory, and they were named just after the Armistice was signed in November 1918. There are Elgin advertisements that date back to 1913 for "The Elgin Wrist Watch for Outdoor Men" and ads from 1915 that refer to "The Outdoor Watch" or "The Strap Watch for Sportsmen," but I do not believe that these were proper given names, only descriptions. Most of these very early Elgin watches were jobber watches, meaning that a watchmaker put them together outside of the factory walls, although in late 1918 Elgin military wristwatches and Elgin military pocket watches were being cased at the factory.

Very few were standardized with very specific cases, dials, and hands—but there were a few. Many variables were possible, such as case style, case size, case material, dial style, dial color, lumed or non-lumed radium hands, semi-hermetic or snap-together cases, crystal guards, straps, and movement grades. If you wanted to design your own wristwatch you would go down to your local jeweler or watchmaker's shop and pick out the options that appealed to you and that would suit your needs. They would then assemble the watch to your specifications. Different shops favored different case manufacturers—such as Philadelphia, Bates & Bacon, Star, Illinois, Wadsworth, Keystone, Fahys, or Crescent—so the case styles would vary from store to store. The shops usually also had preassembled watches from wholesale houses and catalogs, and these were usually in specific styles. This is why you will see many of the same watches that look identical or nearly identical to each other today.

Watches used by the military were a bit different, though. The United States Army had strict specifications for their wristwatches. These specifications were laid out on November 3, 1914, October 24, 1916, and once again sometime in 1918, although that exact date is unknown. This is when the U.S. Army started requiring the dials to be black in color.

Shown opposite are the United States Army October 24, 1916, specifications for wristwatches, as published in the September 26, 1917, issue of *The Horological Review.*

I was unable to find the 1914 specifications, despite a lengthy search through the National Archives in Washington, D.C., and the files at the Combined Arms Research Library at Fort Leavenworth, Kansas. I did, however, find advertisements for Elgin from that time period that show which watches the U.S. Army purchased, so some of those specifications can be deduced from that material.

There has always been a great deal of speculation about "issued" wristwatches in the United States Army during the Great War era. It has long been thought that only officers were wearing wristwatches at this time.

Opposite page:
United States Army specifications for wristwatches, October 24, 1916.

Wrist Watches Purchased for the Army

Chief Signal Officer Supplies Official Specifications for Movements, Cases, and Straps of Wrist Time Pieces Which Government Buys for Soldiers.

WASHINGTON, D. C., Sept. 20.—The Chief Signal Officer of the army has provided your correspondent with the following specifications and drawings of the wrist watch as it is ordered for use by the United States army. It is apparent from the stocks exhibited for sale by many retailers that there is considerable difference between the watches generally sold to soldiers and those the government buys.

The said rules and specifications apply specifically to wrist watches bought by the Government for the United States Army. The question arose as to whether such specifications apply to watches purchased by individual members of the army or presented to the soldiers by their friends. In other words, whether the specifications of the Government in regard to buying watches were, in effect, regulations as to the watches that should be worn.

This question was taken up by THE JEWELER'S CIRCULAR correspondent with officials of the War Department by whom he was informed that the specifications in question apply to *purchases made by the Government* and that there were no restrictions on private purchases made by individual members of the military forces.

This is in line with the ruling of the Adjutant-General's office, published in another column, in regard to tokens carried by soldiers.

The wrist watch specifications adopted Nov. 3, 1914, and revised as of October 24, 1916. The existing specifications and design for strap follows:

GENERAL SPECIFICATION, No. 579-D.
(Oct. 24, 1916.)

WRIST WATCH.

1. *Preliminary.*—This specification describes the general requirements for a wrist watch and wristlet for use by the United States army.

2. *Drawing.*—The following drawing forms a part of this specification:

Strap for wrist watch........1035

3. *Inspection.*—When the order under this specification is complete, the contractor shall notify the Chief Signal Officer of the army, who will cause an inspection to be made. The contractor shall remedy all defects pointed out by the inspector, and will be held accountable for any imperfections which the inspector may overlook.

The watches shall be regulated, and shall be wound and set so that the inspector may readily compare the regulation.

4. *Workmanship and material.*—Workmanship and material throughout shall be first class, and the completed article must be neat in finish and general appearance.

5. *Description of movement.*—(*a*) The movement to be of nickel, standard American manufacture, open face, with lever escapement, compensation balance, Breguet hair spring, nickel damaskeened frames, with not less than seven jewels; diameter of the movement or dial not to be over 1 5/32 inches. It shall be stem winding and stem setting, and shall indicate seconds.

(*b*) *Dial.*—The dial is to be white enamel, and shall have all figures marked in black or blue. A small round spot of radium (luminous) material of sufficient size to be easily seen shall be placed under each set of figures.

(*c*) *Figures.*—Hour figures to be upright Arabic numerals, not less than 1/10 inch in height, and seconds dial orbit shall have each 10-second division indicated in Arabic numerals.

(*d*) *Hands.*—The hands to be of blued steel and of a size and style not confusing to the eye, and shall be coated with a radium (luminous) material.

(*e*) *Case.*—The case shall be of oxidized nickel, with dull, black finish. The case shall be jointed and have inside cap, and the stem shall be practically dust proof. Both the cap and the back of case shall snap well into place, making a good, close joint all around, and shall be easily opened without undue prying.

(*f*) *Crystal.*—The crystal shall be of best quality, clear glass, about 0.06 inch thick, and shall be properly seated in the bezel.

6. *Wristlet.*—A wristlet shall be furnished with each watch. This wristlet shall be made of single-ply genuine pigskin, and shall be of the dimensions and made up as shown upon drawing, paragraph 2(*a*). The cup shall hold the watch tightly, and shall have a snug fitting opening to allow the ring and stem of watch to protrude.

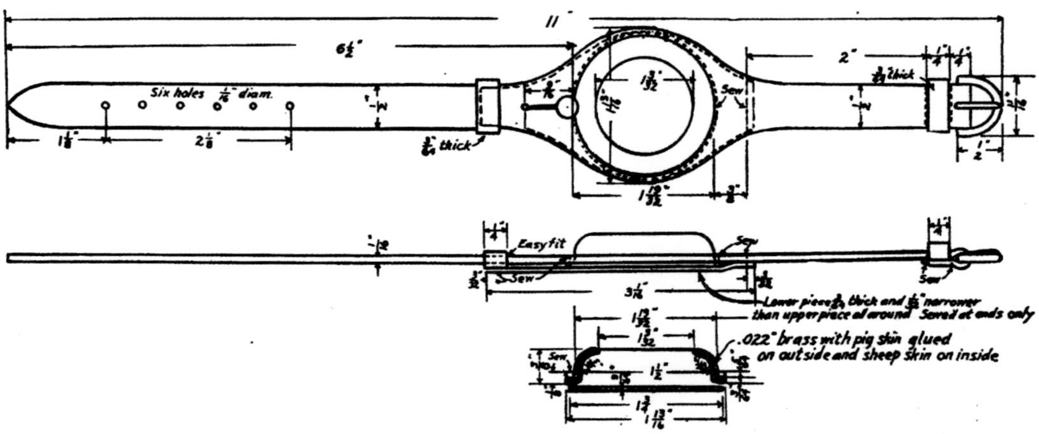

SPECIFICATION OF STRAP FOR WRIST WATCH.
Made of pig skin, tan color. Upper piece, 1/16" thick; lower, 3/64". Upper thread, silk; lower, flax. Buckle brass, black oxidized.

7. *Sample.*—Each bidder shall submit with his proposal, charges prepaid, a sample watch and wristlet of the type he proposes to furnish. The type of movement. case, and quality and workmanship of the wristlet will be considered in awarding the contract.

The sample furnished by the successful bidder will be held for the guidance of the inspector until the final delivery of the order is made.

8. *Guarantee.*—In furnishing watches under this specification, the contractor guarantees for one year after delivery to make repairs to all watches when such repairs are made necessary by defects in material or workmanship, without charge to the Signal Corps.

9. *Packing.*—Great care must be taken in packing material purchased under this specification. Not more than 50 watches shall be packed in one box. Several boxes containing 50 watches each may, however, be inclosed in a larger and stronger box. All boxes shall be marked prominently in red, "KEEP DRY," and shall indicate on the outside the order number, number and kind of instruments contained therein. Articles purchased for export shipment shall be packed with extra care, and a reasonable compensation will be allowed the contractor to cover the expense incident thereto. The contractor will be held accountable for any damage during transportation which could have been avoided by careful packing.

GEORGE P. SCRIVEN,
Brigadier General, Chief Signal Officer of the Army.
Office of the Chief Signal Officer of the Army.

However, I recently found a War Department "General Orders and Bulletins" document, dated February 20, 1913, that states that some enlisted men—privates, corporals, and sergeants—were issued wristwatches. The 1913 document is an update from the original 1911 U.S. Army Drill Regulations Book. It stated that wristwatches were to be issued "previous to drills, inspections and ceremonies when required." However, the United States Army was deployed in Cuba, Panama, the Panama Canal Zone, Haiti, Nicaragua, Mexico, and the Philippines during this time, so, logic leads me to believe that this incredibly important tool was actually being used in the field. If the U.S. Army had watches like this available to them, they would not be used only for drills, as timing is one of the most important factors in any military operation.

GENERAL ORDERS

AND

BULLETINS

WAR DEPARTMENT

1913

WASHINGTON

GOVERNMENT PRINTING OFFICE

1914

	Sema-phore kits.	Field glasses, Type E.	Wrist watches.	Pocket com-passes.
Battery:				
1 sergeant, fifth section		1		
2 corporals, scouts	2	2		
1 corporal, signaler	1	1		
2 privates, signalers	2	2		
3 musicians	3			
Total	8	6		
Battalion headquarters detachment:				
3 corporals, agents	3			
3 corporals, scouts	3	3	3	3
3 privates, scouts	3	3	3	3
1 corporal, signaler	1	1		
1 private, signaler	1			
2 privates, mounted orderlies	2			
3 musicians	3			
Total	16	7	6	6
Regimental headquarters detachment:				
2 sergeants, agents	2			
1 sergeant in charge of scouts	1	1	1	1
3 corporals, scouts	3	3	3	3
3 privates, scouts	3			
1 sergeant, signaler	1	1		
1 private, signaler	1			
5 privates, mounted orderlies	5			
1 trumpeter, sergeant	1			
Total	17	5	4	4
6 batteries	48	36		
2 battalion headquarters	32	14	12	12
Regimental headquarters	17	5	4	4
Total	97	55	16	16

This chart clearly states which enlisted men were to be issued wristwatches. Feb. 20, 1913.

CHAPTER 2

PRICING & RARITY

The pricing for each watch listed in this book is based upon the watch being in overall superb condition. This means the watch will have the original time-period-correct movement with no cosmetic damage and will have been completely overhauled and serviced with a brand new mainspring. The watch must start when wound without being shaken, and the watch must keep proper time over a 24-hour period. The dial must be in good condition without any hairlines, chips, or cracks. The case must be in near-mint condition without any massive damage, like noticeable gold fill or other plating wear spots, including major scratches. Semi-hermetic cases must not have any threading damage; because the layman does not know that these cases are screw on/off, it is common to find pry marks from trying to open the cases the wrong way.

Original factory crowns will always increase a watch's value. Special features of a case, such as a deep-well bezel, swivel lugs, or a coin-edge bezel, will increase the value of a trench watch; wider lug diameter is always desirable, too.

If the watch has been personalized on the case back by the original owner some 90+ years ago, it will not decrease the value. Having the name and hometown of a soldier on the case back is common. Having an original inscription with the name, rank, division, unit, and year gives the watch provenance and proves its historical importance.

Having the correct set of hands is very important and can make or break a watch's value when it comes to the very high-end pieces.

The rarity scale used in this book is based on all of the above factors, including the most important variable: availability. If a watch can be found with only a few clicks of a keyboard, then it is quite common. I use a scale of 1–10. Watches that have very low numbers, such a 1–2, are very common pieces that have major flaws and are not in working order. A 1–2 watch will probably not be running and will have a cracked dial, massive case damage or pitting, and missing parts. I would consider these watches candidates for complete and total restorations or would use them for spare parts. A 3 watch is running but has not been cosmetically restored; some collectors like their watches 100 percent untouched. Watches with a 4–8 rating have been restored and are nice timepieces but are not hard to find. Watches with a 9–10+ rating are top-shelf timepieces that are very rare, with very desirable cases, sizes, movements, dials, and hands. Very few of these watches can be found annually; you might see one or two of them every couple of years, if at all. They are incredibly difficult to obtain, and these watches set the bar.

CHAPTER 3

THE ELGIN NATIONAL WATCH COMPANY

The history of the Elgin National Watch Company dates back to 1864 when it was founded as the National Watch Company by a group of local businessmen looking to compete with the American Watch Company in Waltham, Massachusetts. Some of the key players involved with this venture were John C. Adams, Benjamin W. Raymond, Howard Culver, Joseph Pyerson, George Wheeler, and Philo Carpenter, along with nine Chicago-based investors. Many of these men subsequently had watch movements named for them. A few streets in Elgin, Illinois, bear their names as well. The company was organized on August 18, 1864, and shortly thereafter several employees were hired away from Waltham. These men became known as the "Seven Stars." By 1866, the factory was up and running, and the first watch was sold in April of 1867, a full-plate, size 18s movement with the serial number 101. The name was officially changed to the Elgin National Watch Company in 1874 as the name "Elgin" had become the common name for their watches. By 1877, the company had already produced more than 500,000 movements; by March of 1882, that number rose to 1,000,000 total movements. By the turn of the century, Elgin employed nearly 3,000 men and women.

During World War I, Elgin produced approximately 1,000,000 movements per year of various sizes. Many of the movements being manufactured were the 3/0s, 0s, and 6s that were used in military watches. The numbers indicate the size. On average, a size 3/0s trench watch will measure about 30.5 mm–31.5 mm. A size 0s will measure about 31.5 mm–34.5 mm. The giant size 6s will measure between 37 mm and 40 mm. These measurements are going across the case, without including the crown.

High production numbers continued through the 1920s, but then production was cut by over half when the Great Depression began in late 1929, and the production numbers stayed at that lower rate for several years thereafter.

During World War II, Elgin started manufacturing items for the war effort, such as fuses for artillery shells, aviation clocks, stopwatches, navigation watches, aircraft compasses, and altimeters. Elgin produced more than 1,000,000 military wristwatches during WWII.

In September of 1951, the company manufactured its 50,000,000th watch movement; they were the very first jeweled watch company in the world to reach that figure. It was a 21 jewel Lord Elgin in a gold case.

The Elgin National Watch Company factory in its glory days. *Photo courtesy of the Elgin History Museum.*

This sign now stands where the Elgin National Watch Company factory was located.

But trouble was on the horizon. Swiss watches began outselling American watches by 4 to 1 by the mid-1950s. In 1958, Elgin posted a loss of over two million dollars, and in 1960, the Elgin Watchmaker's College closed their doors for good. The very last Elgin movement made in the United States was produced at their South Carolina plant in January 1968.

In 1966, the original manufacturing plant in Elgin, Illinois, was demolished, and on October 2, 1966, the Elgin clock tower was destroyed by dynamite. The bell from that tower now resides in front of the Elgin Historical Museum building on Park Street for all to visit.

A view from across the Fox River to where the Elgin factory once stood.

The "Old Main" is where the Elgin Historical Museum is located. If you are ever in the Elgin, Illinois area, stop in for a visit.

The Elgin National Watch Company factory bell now resides at the Historical Museum.

CHAPTER 4

THE STAR DIAL SERIES

Today, there is much speculation about the Black Star Dial Elgin trench watches: were they actually government-issued wristwatches, or were they simply sold at Post Exchanges? The answer is that they were both. The United States Army made a purchase of Black Star Dial wristwatches around July 1918, although the exact number bought is unknown. These watches were also available for sale at post exchanges at military bases for private purchases, although the specs were different. The PX watches did not have a standardized case, so many different case styles were available in options including silverode, gold-filled, nickel, sterling silver, nickel silver, and, very rarely, solid gold. The U.S. Army-issued Elgins had solid nickel, semi-hermetic cases made by the Illinois Watch Case Company (also located in Elgin) and Joseph Fahys & Company (located in New York City).

These Army-issued watches did not have any government or Ordnance Department (ORD) markings when they were first purchased; this practice was not started for many more years. But between WWI and WWII, when an Elgin wristwatch went in for servicing, the Army watchmaker would then stamp the outside of the case back with the appropriate ORD marks for that watch.

Per U.S. Army specifications, the watch case had to be painted flat black and the case had to be semi-hermetic. The hour and minute hands had to be luminous, along with the hour markers.

Some of the U.S. Army-issued Elgin Black Star Dial wristwatches bought by the government served United States forces in both WWI and WWII; a few of these

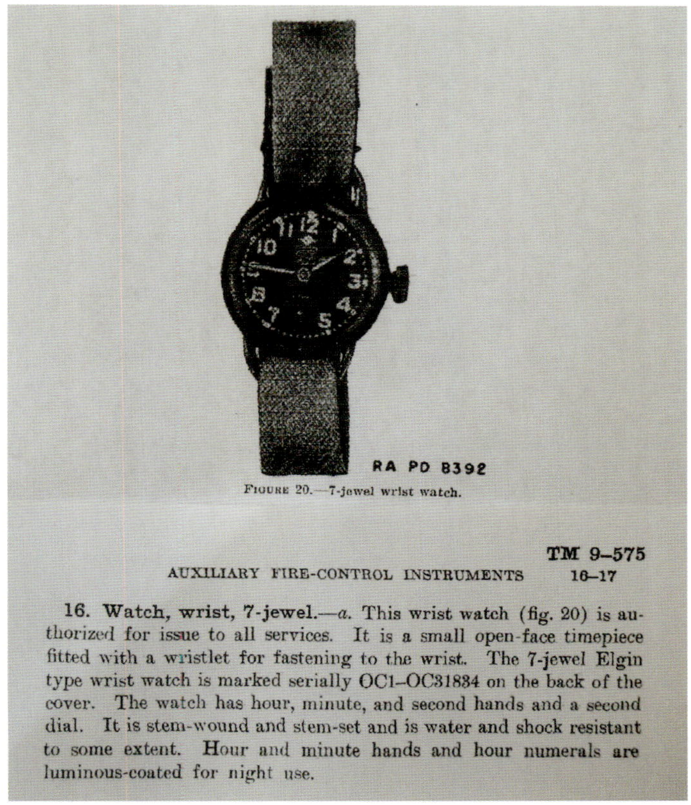
RA PD 8392

FIGURE 20.—7-jewel wrist watch.

TM 9–575

AUXILIARY FIRE-CONTROL INSTRUMENTS 16–17

16. Watch, wrist, 7-jewel.—*a.* This wrist watch (fig. 20) is authorized for issue to all services. It is a small open-face timepiece fitted with a wristlet for fastening to the wrist. The 7-jewel Elgin type wrist watch is marked serially OC1–OC31834 on the back of the cover. The watch has hour, minute, and second hands and a second dial. It is stem-wound and stem-set and is water and shock resistant to some extent. Hour and minute hands and hour numerals are luminous-coated for night use.

The U.S. Army's 1942 War Department Technical Manual TM 9-57 shows an Elgin Black Star Dial. Notice the tapered lugs on the solid nickel semi-hermetic Illinois case. They start out at 15.5 mm then go down to 14 mm for the band. The crystal size on this case is much bigger than what you would normally see on a size 3/0s case; it measures 27.8mm.

watches had a service life of at least twenty-four years. Evidence of this can found in the War Department Technical Manual TM 9-575, which is the manual for auxiliary fire control instruments (field glasses, eyeglasses, telescopes, and watches) dated May 4, 1942. On pages 20 and 21 of that manual is a picture of an Elgin wristwatch and description of the Black Star Dial, indicating the wristwatch was still in use during WWII.

Obtaining one of these genuine WWI U.S. Army-issued Elgins today is extremely difficult. In all of my years of watchmaking and collecting, I have only seen five Black Star Dial Elgin watches that had the ORD markings on the outside of the case back, making them very rare timepieces. If you are on the hunt for a Black Star Dial Elgin, you can probably find them on internet auction sites, but most likely they will be lacking the correct and authentic ORD markings—they will just be the very common unmarked versions.

This is a 1918 Black Star Dial Elgin Trench Watch that was U.S. Army-issued with the correct and genuine ORD markings.

Far left:
Once the bezel and case back have been unscrewed, you can still see the remnants of the required original flat black paint.

Left:
Removing the case back reveals the absolutely beautiful and original grade 462 movement with 7 jewels, as well as the original factory crown.

1918 ELGIN, BLACK STAR DIAL

When I first found this watch in February 2012 at a National Association of Watch & Clock Collectors regional event in Florida, it was not working and in need of many repairs. It needed a new hairspring and a new mainspring, and the metal bushings also needed to be polished. A correct glass crystal had to be installed, and the watch needed a proper cleaning and timing adjustment. Only mechanical repairs have been done—nothing cosmetic, other than a correct glass crystal and a new pigskin band. Fully restoring this U.S. Army-issued Elgin would bring down its value drastically and ruin the watch. This ninety-six-year-old watch is in remarkable original condition for its age.

1918 Elgin, Black Star Dial, replacement skeleton hands installed by a U.S. Army watchmaker, hour marker lum dots drilled by a U.S. Army watchmaker (originally the numerals were covered with radium), Illinois Watch Case Company solid nickel semi-hermetic case, original factory crown, original and genuine ORD markings on the case back, unrestored and unpolished, date stamped original enamel dial, size 3/0s, 7 jewels, grade 462 movement, serial number 21080456, 14mm genuine pigskin band.

Value: $2,000–$2,500
Rarity: 10+

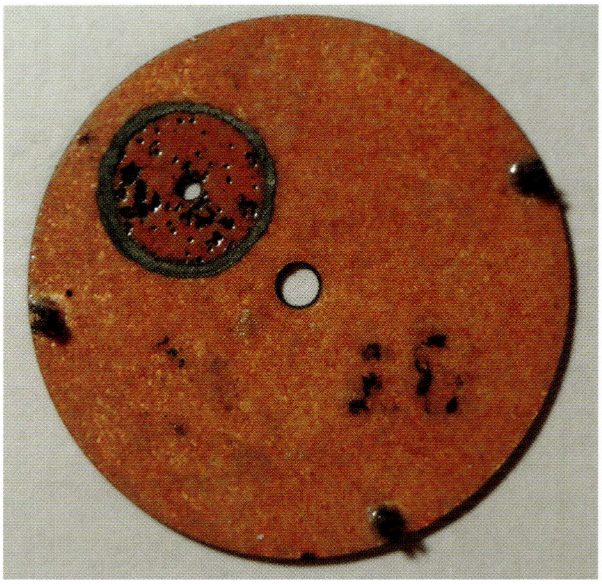

You can still see the very faint remnants of the Elgin factory date-stamp on the back side of this enamel dial. The month is not legible, but you can still see what's left of the "18" (for 1918) on the right-hand side.

Semi-hermetic IWC Co case with remnants of the original black finish.

ELGIN, ENAMEL WHITE STAR DIAL

This exact semi-hermetic, solid nickel case style, from the Illinois Watch Case Company, with 14mm lugs was also available with an enamel White Star Dial. The case back has not been stamped with ORD markings, but the case was originally painted all black. Thick blue steel hands, BOLD Arabic numerals, 48 black stars, sub-second hand, Eagle crest, size 3/0s, grade 462, serial number 21609072, 14mm genuine Pigskin strap.

Value: $1,500–$1,700
Rarity: 10

Above: Notches for the chuck key are on the bezel near the 2 and 8 o'clock positions. Middle right: Note the screw-down crown.

In June 1919, when Jacques Depollier & Son came out with a waterproof and dustproof watch case, the Great War had officially not ended yet. The Armistice that was signed on November 11, 1918, had stopped the fighting, but the Treaty of Versailles wasn't signed until June 28, 1919. This watch never saw any action during the fighting of the Great War, but it is still considered a WWI timepiece because it was available before the Treaty of Versailles was signed. Depollier & Son had a very close relationship with the Waltham Watch Company (all of their advertising from this period only features Waltham movements), but Depollier cases were also used with Elgin and Illinois movements. In mid-1919 the U.S. Army bought a quantity of these Depollier waterproof cases and installed movements and military dials from Elgin and Illinois that they already had on hand. Some movements and dials from Elgin and Illinois that were currently cased in nonwaterproof cases were also re-cased into the Depollier cases. For these watches, the movement serial number pre-dates the case by one to three years. These watches are considered among the most collectible from the Great War era. They measure 42mm lug to lug and 42mm including the crown.

DEPOLLIER & SON WATERPROOF AND DUSTPROOF STRAP WATCH, ELGIN MOVEMENT, BLACK STAR DIAL

Most of the original flat black paint on the case of the watch shown here is still intact, increasing the watch's value. Look at the outside of the case back: notice the circular groove? When this watch was brand new, it had a 14K solid gold disk in this groove. According to Depollier, this heat-insulating disk protected the delicate movements from body heat, which had a tendency to gum up the watch oil. Watches with the 14K solid gold disk still installed are rare finds today. Look at the top of the watch, and notice the two notches: one is between the 1 and 2, the other is between the 7 and 8. This is for a chuck key that opens and closes the threaded hermetic case. These notches are found on the outside of the case back as well.

Value: $1,500–$2,000
$2,500+ if the 14K solid gold disk is with the watch.
Rarity: 10+

An ad for the Depollier Waterproof and Dustproof Watch from Oct. 1, 1919.

Waltham Movement

Prestige Accuracy

The Depollier Waterproof and Dustproof Strap Watch

FIELD AND MARINE

Registered U. S. Pat. Off.

The Depollier Waterproof Case now adopted by the Signal Corps of the United States Army for the saving of watch movements purchased during the war and for future use. Ordinary strap watches were not constructed to withstand the wear and tear of field duty.

A heat-insulated disk protects the delicate movement from the injurious body heat of the arm, which has a tendency to dry or gum up the watch oil.

Waterproof Oxidized Case with 14-k Solid Gold Disk on Back and 15-J Waltham Movement - - $45.00

Patented

Write for Booklet

JACQUES DEPOLLIER & SON

Manufacturers of High Class Specialties for Waltham Watches

15 Maiden Lane - New York, N. Y.

Dubois Watch Case Company, Established 1877

Stamped U. S. A. only for the U. S. Army

The Literary Digest for February 22, 1919 101

—and for the Traveling Man

Released by the Armistice, the three most popular styles are now available for civilian use. Let your Jeweler show you these Elgins.

"Pershing"

This is the "cushion shape," as pictured in above close-up of the Big Fist. Heavy pigskin strap.
Nickel - - - - $21
Gold-filled or Silver - - - 25

"Kitchener"

"Foch"

The "Foch" style is the round shape as here shown, with khaki or pigskin strap.
Nickel - - $18.50
Gold-filled or Silver - $21.00

The "Kitchener" style is the square shape as shown above. Wide pigskin strap, double.
Nickel - - - $20
Gold-filled or Silver - - $25

Remember, these strap watches are but three of the many Elgin styles

Elgin Strap Watch

"Life," says the traveling man, "is just catching one train after another. Like any other successful business, mine has to run on schedule, and my Elgin strap watch is the answer. With the time always before me, I know exactly where I'm at.

"Yes, and it's a satisfaction to own a watch that's never dropped, never left under my pillow—as plainly visible in my berth at midnight as at high noon. For a handy *extra* watch for a man on the road—or for any other active man—there's nothing like the Elgin strap watch with the black dial."

Genuine black-dial full-luminous Elgins, in your choice of the three styles of case and strap here pictured—now ready at your Jeweler's. The Armistice releases them for civilian use.

Handy in winter, when gloves and overcoat make fumbling in your pocket clumsy and unsafe.

Handy in summer, when you wear no vest and half the time no coat.

Buy one while you can get it.

All three styles have genuine Elgin movements—American through and through. Specially constructed for hard outdoor use. Wear one a week and you'd feel lost without—

"The Handy Extra Watch for Men"

After the Armistice was signed on November 11, 1918, essentially ending The Great War, these watches were released to the public for civilian use, according to the Elgin ad that appeared in *The Literary Digest* magazine on February 22, 1919. It was quite common to wear a wristwatch on the inside of your wrist back then. An identical ad also appeared in the March 8, 1919, edition of *The Saturday Evening Post*.

Shown with this 1919 Elgin Strap Watch ad are examples of the three watches advertised. They were named for the three main victorious leaders of the Allied Forces during the war: the Pershing (United States), the Foch (France), and the Kitchener (England). These are the very first "named" men's Elgin wristwatches. These watches did *not* have these names during the Great War. It's a very common misconception that all "Black Star Dial" Elgin's are called Pershings. This is simply *not* the case.

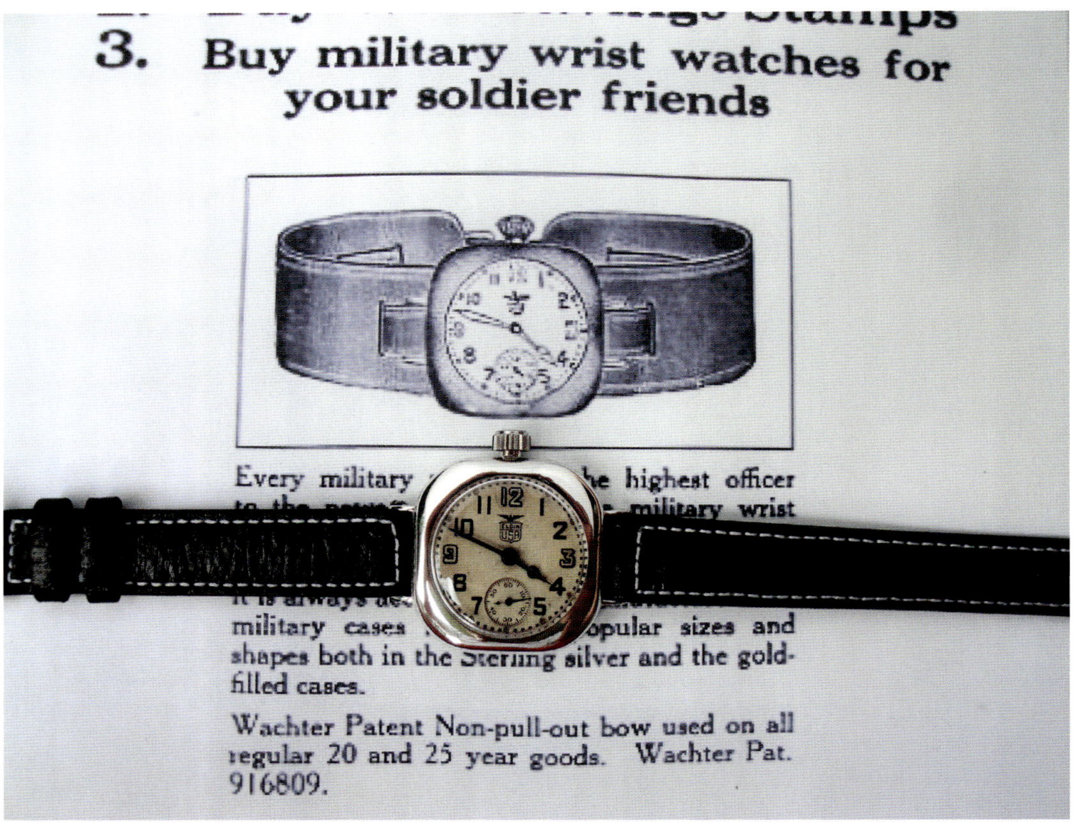

Here is another wonderful ad that actually shows the crown in the 12 o'clock position from the Star Watch Case Company. Only three ads are known to exist that show an Elgin White Star Dial; three are known to exist that show Elgin's Black Star Dials as well. When worn on the inside of the wrist, the numerals would be straight up and down.

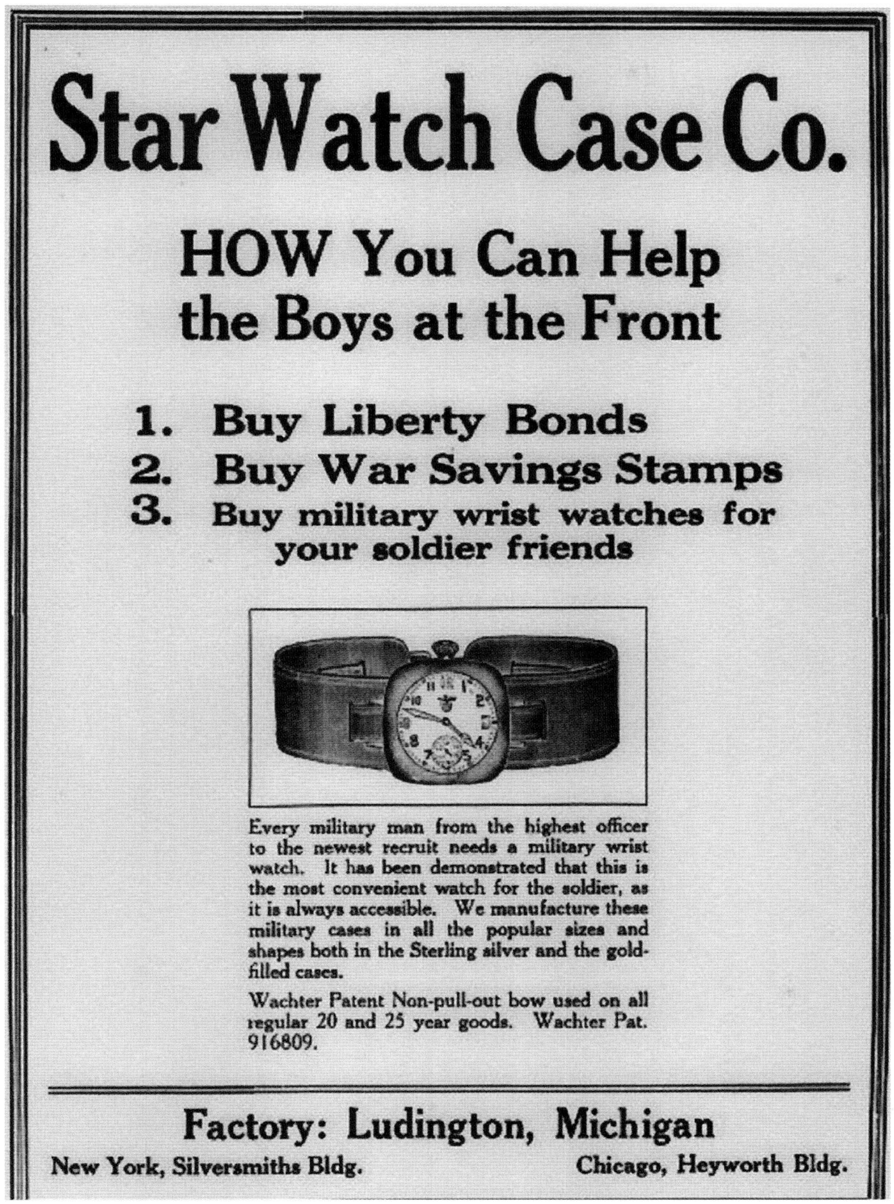

Here is the full ad from the Star Watch Case Company that was featured in the October 30, 1918, edition of *The Jewelers' Circular.*

Page 62 THE KEYSTONE December, 1918

STAR WATCH CASE COMPANY

Peace on Earth, Good Will to Men

The nightmare of War is past and we awaken to a new dawn of prosperity.

This will be an age of Physical fitness. Our boys have set the pace and we must live up to it. Sports and athletics of all kinds will be the vogue and there will be an increased demand for men's and women's wrist watches. These have proved themselves to be an essential to all devotees of sport. We manufacture these in all sizes and popular shapes, both for men and women.

Cut shows one of our men's wrist watches.

Safety Bow Licensed Under Wachter Pat. 916809

FACTORY, LUDINGTON, MICH.

NEW YORK, Silversmiths Building
CHICAGO, Heyworth Building
SAN FRANCISCO, Mutual Bank Bldg.

In this advertisement from the Star Watch Case Company, only two months later than the October one, the message has changed now that the Armistice has been signed. Star is promoting "Peace on Earth, Good Will to Men" instead of asking you to buy war bonds or stamps.

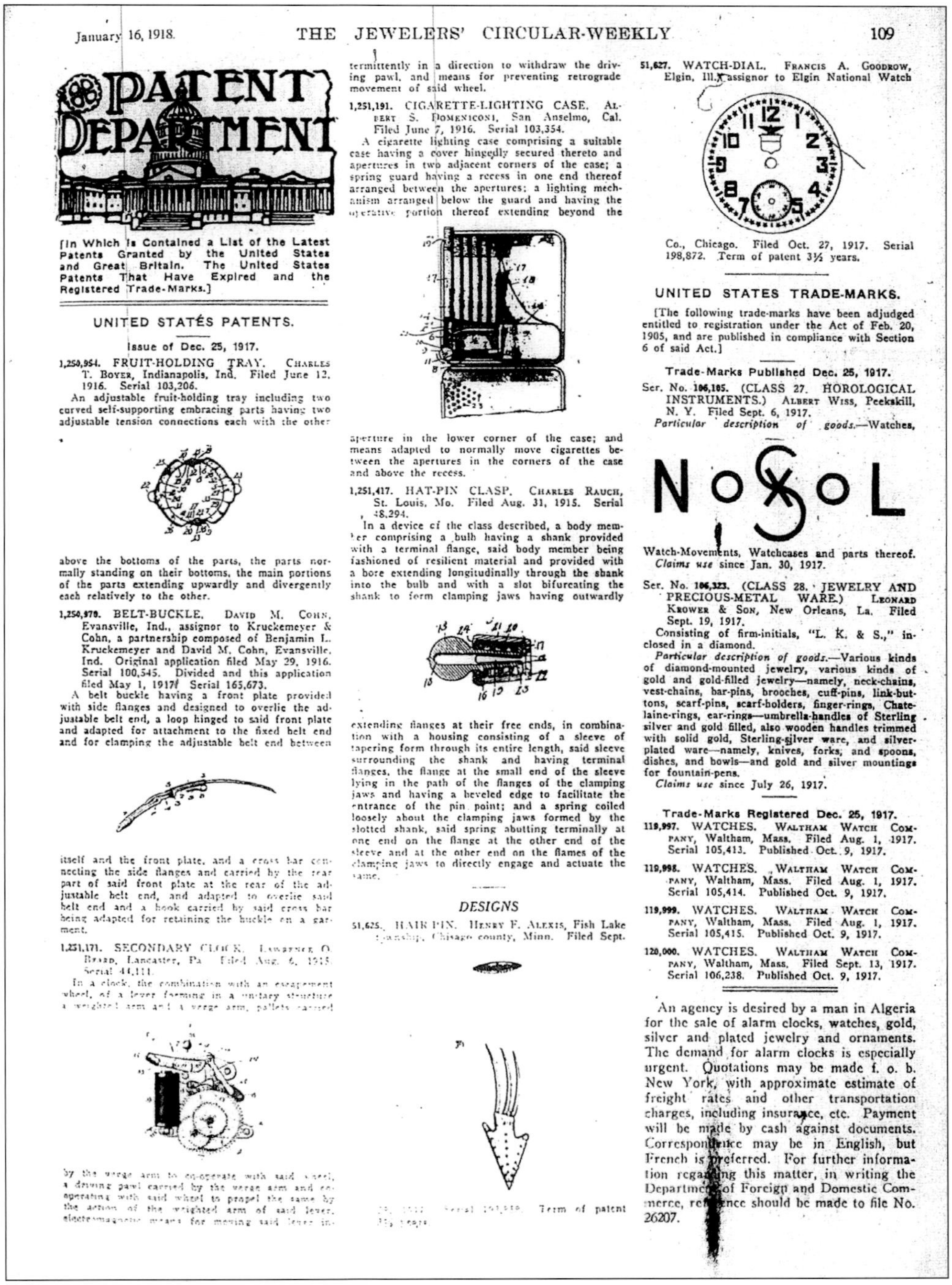

January 16, 1918. THE JEWELERS' CIRCULAR-WEEKLY 109

PATENT DEPARTMENT

[In Which Is Contained a List of the Latest Patents Granted by the United States and Great Britain. The United States Patents That Have Expired and the Registered Trade-Marks.]

UNITED STATES PATENTS.

Issue of Dec. 25, 1917.

1,250,954. FRUIT-HOLDING TRAY. CHARLES T. BOYER, Indianapolis, Ind. Filed June 12, 1916. Serial 103,206.
An adjustable fruit-holding tray including two curved self-supporting embracing parts having two adjustable tension connections each with the other above the bottoms of the parts, the parts normally standing on their bottoms, the main portions of the parts extending upwardly and divergently each relatively to the other.

1,250,970. BELT-BUCKLE. DAVID M. COHN, Evansville, Ind., assignor to Kruckemeyer & Cohn, a partnership composed of Benjamin L. Kruckemeyer and David M. Cohn, Evansville, Ind. Original application filed May 29, 1916. Serial 100,545. Divided and this application filed May 1, 1917. Serial 165,673.
A belt buckle having a front plate provided with side flanges and designed to overlie the adjustable belt end, a loop hinged to said front plate and adapted for attachment to the fixed belt end and for clamping the adjustable belt end between itself and the front plate, and a cross bar connecting the side flanges and carried by the rear part of said front plate at the rear of the adjustable belt end, and adapted to overlie said belt end and a hook carried by said cross bar being adapted for retaining the buckle on a garment.

1,251,171. SECONDARY CLOCK. LAWRENCE O. BRAND, Lancaster, Pa. Filed Aug. 6, 1915. Serial 44,111.
In a clock, the combination with an escapement wheel, of a lever forming in a unitary structure a weighted arm and a verge arm, pallets carried by the verge arm to co-operate with said wheel, a driving pawl carried by the verge arm and co-operating with said wheel to propel the same by the action of the weighted arm of said lever, electromagnetic means for moving said lever intermittently in a direction to withdraw the driving pawl, and means for preventing retrograde movement of said wheel.

1,251,191. CIGARETTE-LIGHTING CASE. ALBERT S. DOMENICONI, San Anselmo, Cal. Filed June 7, 1916. Serial 103,354.
A cigarette lighting case comprising a suitable case having a cover hingedly secured thereto and apertures in two adjacent corners of the case; a spring guard having a recess in one end thereof arranged between the apertures; a lighting mechanism arranged below the guard and having the operative portion thereof extending beyond the aperture in the lower corner of the case; and means adapted to normally move cigarettes between the apertures in the corners of the case and above the recess.

1,251,417. HAT-PIN CLASP. CHARLES RAUCH, St. Louis, Mo. Filed Aug. 31, 1915. Serial 48,294.
In a device of the class described, a body member comprising a bulb having a shank provided with a terminal flange, said body member being fashioned of resilient material and provided with a bore extending longitudinally through the shank into the bulb and with a slot bifurcating the shank to form clamping jaws having outwardly extending flanges at their free ends, in combination with a housing consisting of a sleeve of tapering form through its entire length, said sleeve surrounding the shank and having terminal flanges, the flange at the small end of the sleeve lying in the path of the flanges of the clamping jaws and having a beveled edge to facilitate the entrance of the pin point; and a spring coiled loosely about the clamping jaws formed by the slotted shank, said spring abutting terminally at one end on the flange at the other end of the sleeve and at the other end on the flames of the clamping jaws to directly engage and actuate the same.

DESIGNS

51,625. HAIR PIN. HENRY F. ALEXIS, Fish Lake Township, Chisago county, Minn. Filed Sept. [illegible] Serial [illegible] Term of patent 3½ years.

51,627. WATCH-DIAL. FRANCIS A. GOODROW, Elgin, Ill., assignor to Elgin National Watch Co., Chicago. Filed Oct. 27, 1917. Serial 198,872. Term of patent 3½ years.

UNITED STATES TRADE-MARKS.

[The following trade-marks have been adjudged entitled to registration under the Act of Feb. 20, 1905, and are published in compliance with Section 6 of said Act.]

Trade-Marks Published Dec. 25, 1917.

Ser. No. 106,105. (CLASS 27. HOROLOGICAL INSTRUMENTS.) ALBERT WISS, Peekskill, N. Y. Filed Sept. 6, 1917.
Particular description of goods.—Watches,

NOSOL

Watch-Movements, Watchcases and parts thereof.
Claims use since Jan. 30, 1917.

Ser. No. 106,323. (CLASS 28. JEWELRY AND PRECIOUS-METAL WARE.) LEONARD KROWER & SON, New Orleans, La. Filed Sept. 19, 1917.
Consisting of firm-initials, "L. K. & S.," inclosed in a diamond.
Particular description of goods.—Various kinds of diamond-mounted jewelry, various kinds of gold and gold-filled jewelry—namely, neck-chains, vest-chains, bar-pins, brooches, cuff-pins, link-buttons, scarf-pins, scarf-holders, finger-rings, Chatelaine-rings, ear-rings—umbrella-handles of Sterling silver and gold filled, also wooden handles trimmed with solid gold, Sterling-silver ware, and silver-plated ware—namely, knives, forks, and spoons, dishes, and bowls—and gold and silver mountings for fountain-pens.
Claims use since July 26, 1917.

Trade-Marks Registered Dec. 25, 1917.

119,997. WATCHES. WALTHAM WATCH COMPANY, Waltham, Mass. Filed Aug. 1, 1917. Serial 105,413. Published Oct. 9, 1917.

119,998. WATCHES. WALTHAM WATCH COMPANY, Waltham, Mass. Filed Aug. 1, 1917. Serial 105,414. Published Oct. 9, 1917.

119,999. WATCHES. WALTHAM WATCH COMPANY, Waltham, Mass. Filed Aug. 1, 1917. Serial 105,415. Published Oct. 9, 1917.

120,000. WATCHES. WALTHAM WATCH COMPANY, Waltham, Mass. Filed Sept. 13, 1917. Serial 106,238. Published Oct. 9, 1917.

An agency is desired by a man in Algeria for the sale of alarm clocks, watches, gold, silver and plated jewelry and ornaments. The demand for alarm clocks is especially urgent. Quotations may be made f. o. b. New York, with approximate estimate of freight rates and other transportation charges, including insurance, etc. Payment will be made by cash against documents. Correspondence may be in English, but French is preferred. For further information regarding this matter, in writing the Department of Foreign and Domestic Commerce, reference should be made to file No. 26207.

At top right on this page from *The Jewelers' Circular* is information about the patent for the Star Dial Elgins.

When I first started buying and collecting WWI American trench watches, I was immediately drawn to the Star Dial Elgins because of the beauty and intricate detail of the dials. They are some of the most iconic and recognizable WWI American trench watches. But when I first started researching them, I found more speculation than facts. These military dials was never listed in the Elgin material catalogs, and all of the hard copy records have been lost in time. In October 2010, though, I was searching through old copies of *The Jewelers' Circular*, a weekly trade magazine for jewelers and watchmakers of that era, and I discovered a list of patents that had been recently granted within the trade. After years of searching I finally had a lead! This clue lead me to the United States Patent and Trademark Office. Now that I had a name and a date for the patent, the answers starting flowing. The "Star Dial" series designer's name was Francis A. Goodrow; he was an employee of the Elgin National Watch Company for fifty-two years and foreman of the engraving department. Here was my first clue that appeared in the January 16, 1918 edition.

Elgin did not waste any time going into production of the dials once the patent application paperwork was filed. The very first dials were available in November 1917. The patent was actually granted for three and a half years on December 25, 1917. I believe that the two metal White Star Dial designs were available first, followed by the enamel thin-numeral Black Star Dial, and then, several months later, the enamel thick-numeral Black Star Dial. Case styles varied; the dates given are for the dials only. (Note: I have seen a massive influx of fake reproduction dials in recent months; be sure when adding one of these watches to your collection that it's the genuine article!)

October 1917, 1st design

October 1917, 2nd design

January 1918, 3rd design

July 1918, 4th design

In late 2012, I came across a bit of an oddity: an *enamel* White Star Dial that never had luminescent material on the dial or hands. The other two White Star Dial designs are *metal* dials. This dial features BOLD Arabic numerals and a circle at the five minute intervals, slightly different than the other four designs. The Eagle Crest is identical with the ELGIN USA. I did multiple tests, including ultraviolet, ultrasonic, weight and measurements, on this enamel dial to check for authenticity. The numerals and figures were fired into the dial, just as they should have been. The back side of the dial is exactly as it should have been: orange with slight black and red spotting. This is possibly a prototype dial that was never mass produced, as so few are known to exist.

Top right: Back side of the enamel White Start Dial. **Above:** Mid to late 1918, 5th design. The is an Elgin White Star Dial in an Illinois "General Funston" case.

1918 ELGIN, BLACK STAR DIAL

This trench watch is one of the finest watches featured in this book; it simply has the best of everything available in 1918. It is very uncommon to find a Great War Elgin trench watch with a 14K solid gold case, especially with this very rare case design. I consulted with several of the most knowledgeable collectors in the country, and none of us had seen an American-made case like this before. The case was manufactured by the Keystone Watch Case Company, and it has a really unique feature: both the bezel and case back are hinged. It has the very desirable 16mm wide lug configuration. The case alone without the movement or crystal weighs in at 11.2 grams, a few grams heavier than other 14K solid gold American-made cases that I've seen. It still has the original 14K solid gold factory crown, which greatly increases the watch's value. Grade 464, size 3/0s, 15 jewels, serial number 21018228. This was the best size 3/0s movement Elgin produced in 1918. The original blue steel hands have been relumed (luminescent material reapplied) with tight crisp edges. New inner sleeve, new mainspring, new glass crystal and a new 16mm pigskin strap.

Value: $3,000–$3,500
Rarity: 10+

Top right: Hinged bezel. **Bottom right:** Hinged case back. **Bottom right:** Hinged case back revealing grade 464 movement with 15 jewels.

Top: Size 3/0s Keystone case clearly marked for 14K solid gold content.
Middle: 16mm wide lugs.
Bottom: The lugs are rather stout, measuring 2.25 mm thick.

DESIGN.

F. A. GOODROW.

WATCH DIAL.

APPLICATION FILED OCT. 27, 1917.

51,627. Patented Dec. 25, 1917.

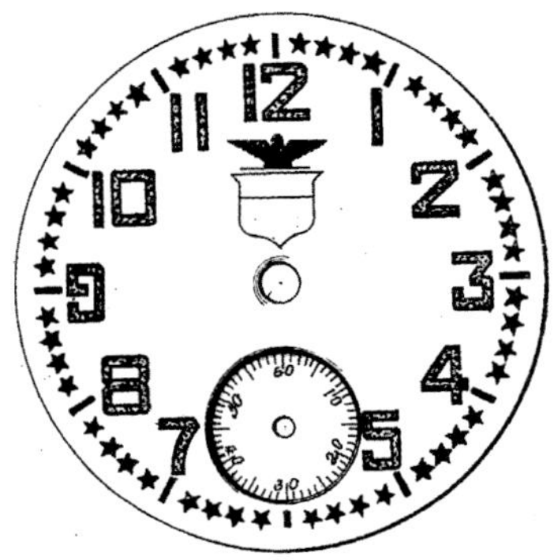

Inventor

Francis A. Goodrow

Witnesses

Floyd R. Cornwell

Grace P. Brereton

By Sturtevant & Mason

Attorneys

UNITED STATES PATENT OFFICE.

FRANCIS A. GOODROW, OF ELGIN, ILLINOIS, ASSIGNOR TO ELGIN NATIONAL WATCH COMPANY, OF CHICAGO, ILLINOIS, A CORPORATION OF ILLINOIS.

DESIGN FOR A WATCH-DIAL.

51,627. **Specification for Design.** **Patented Dec. 25, 1917.**

Application filed October 27, 1917. Serial No. 198,872. Term of patent 3½ years.

To all whom it may concern:

Be it known that I, Francis A. Goodrow, a citizen of the United States, residing at Elgin, in the county of Kane and State of Illinois, have invented a new, original, and ornamental Design for Watch-Dials, of which the following is a specification, reference being had to the accompanying drawing, forming part thereof.

In the drawing,

The figure is a plan view of a watch dial embodying my design.

I claim:

The ornamental design for a watch dial, as shown.

FRANCIS A. GOODROW.

Copies of this patent may be obtained for five cents each, by addressing the "Commissioner of Patents, Washington, D. C."

Francis Goodrow's 1917 patent application for Elgin's Star Dial design.

1918 ELGIN, BLACK STAR DIAL

Enameled center reservoir military hands, 9K solid gold Keystone Watch Case Company case with a hinged case back, replacement crown, size 3/0s, 15 jewels, 3-piece 10mm Bund Strap.

Value: $1,000–$1,200
Rarity: 10

1918 ELGIN, BLACK STAR DIAL

Enameled center reservoir military hands, Philadelphia semi-hermetic sterling silver case with deep-well bezel, sterling silver factory crown, size 3/0s, 7 jewels, 1 piece 16mm nylon strap.

Value: $600–$750
Rarity: 8

1918 ELGIN, PERSHING, BLACK STAR DIAL

Relumed, skeleton hands, Star Watch Case Company nickel cushion case, factory crown, size 3/0s, 7 jewels, 19mm pigskin strap.

Value: $400–$500
Rarity: 6

1918 ELGIN, BLACK STAR DIAL

Relumed center reservoir military hands, 14K sold gold Illinois Watch Case Company case with a hinged case back, factory crown, 3/0s, 7 jewels, 14mm pigskin strap. (8.8 grams of 14K gold)

Value: $1,700–$2,000
Rarity: 10

1918 ELGIN, BLACK STAR DIAL

Relumed center reservoir military hands, Illinois Watch Case Company nickel semi-hermetic case, replacement crown, size 3/0s, 15 jewels, 1-piece 16mm nylon strap.

Value: $900–$1,200
Rarity: 9

1918 ELGIN, BLACK STAR DIAL

Enameled center reservoir military hands, Philadelphia silverode case with deep-well bezel and an offset crown, factory crown, size 3/0s, 7 jewels, 16mm pigskin band.

Value: $500–$600
Rarity: 7

1918 ELGIN, ENAMEL WHITE STAR DIAL

Arabic numerals, blue steel hands, Illinois Watch Case Company sterling silver case, swivel lugs, hinged case back, factory crown, size 3/0s, 7 jewels, new hand-sewn olive drab webbing strap with original 1918 date-stamped, sterling silver strap hardware.

Value: $1,000–$1,500
Rarity: 10

1918 ELGIN, BLACK STAR DIAL

Relumed center reservoir military hands, Illinois Watch Case Company sterling silver semi-hermetic case with swivel lugs, deep-well bezel, replacement crown, size 3/0s, 7 jewels, 14mm pigskin band.

Value: $700–$900
Rarity: 8

1918 ELGIN, BLACK STAR DIAL

Relumed center reservoir military hands, "no-name" nickel case made by Wadsworth, replacement crown, size 3/0s, 7 jewels, 1-piece 16mm nylon strap.

Value: $500–$600
Rarity: 7

1918 ELGIN, BLACK STAR DIAL

Relumed center reservoir military hands, Illinois Watch Case Company sterling silver case, factory crown, size 3/0s, 15 jewels, 1 piece 14mm nylon strap.

Value: $500–$600
Rarity: 7

1918 ELGIN, BLACK STAR DIAL

Enameled center reservoir military hands, 10K gold-filled Star Watch Case Company case with a hinged case back, replacement crown, 3/0s, 15 jewels, 3-piece 10mm bund strap.

Value: $500–$600
Rarity: 7

1918 ELGIN, BLACK STAR DIAL

Relumed center reservoir military hands, correct silver second hand, Joseph Fahys & Company oresilver semi-hermetic case, factory crown, size 3/0s, 15 jewels, 16mm pigskin leather band.

Value: $550–$650
Rarity: 7

1917 ELGIN, WHITE STAR DIAL

Metal dial, center reservoir military hands, Philadelphia silverode case with an offset crown, factory crown, size 3/0s, 7 jewels, 3-piece 10mm bund strap.

Value: $600–$800
Rarity: 8

1917 ELGIN, WHITE STAR DIAL

Metal dial, center reservoir military hands, Philadelphia silverode case with an offset crown and a deep-well bezel, replacement crown, size 3/0s, 7 jewels, 1 piece 16mm nylon strap.

Value: $550–$650
Rarity: 7

1917 ELGIN, WHITE STAR DIAL

Metal dial, center reservoir military hands, Philadelphia sterling silver semi-hermetic case with an offset crown, factory crown, size 3/0s, 7 jewels.

Value: $700–$800
Rarity: 8

1918 ELGIN, ENAMEL WHITE STAR DIAL

Blue steel hands, Illinois Watch Case Company sterling silver barrel case, 23.5 grams of sterling silver, factory crown, size 3/0s, 15 jewels, 20mm genuine pigskin strap.

Value: $1,500–$2,000
Rarity: 10+

1918 ELGIN, BLACK STAR DIAL

Enameled center reservoir military hands, Illinois Watch Case Company 10K gold- filled case, factory gold-filled crown, replacement 1915 movement, size 3/0s 15 jewels, 14mm pigskin band.

Value: $600–$800
Rarity: 7

1918 ELGIN, BLACK STAR DIAL

Enameled center reservoir military hands, Philadelphia silverode case with an offset crown, factory crown, size 3/0s, 7 jewels.

Value: $500–$600
Rarity: 7

1918 ELGIN, BLACK STAR DIAL

Enameled center reservoir military hands, Illinois Watch Case Company sterling silver case, crown at 12, replacement crown, size 3/0s, 7 jewels, 14mm pigskin band.

Value: $500–$600
Rarity: 7

OCTOBER 1918, BLACK STAR DIAL

Perfect dial with no cracks, perfectly defined eagle wings (the wings are very rarely this crisp), date-stamped on back side.

Dial-only value: $100
Dial value with slightly blurry wings: $50
Rarity: 10

OCTOBER 1917, WHITE STAR DIAL

Metal dial, exaggerated 12 – 3 – 9 for lum material, regular 1-2-4-5-7-8-10-11, factory drilled lum holes.

This condition or better: $100
Poorly aged dial with damage: $50
Rarity: 10

CHAPTER 5

THE GIANT SIZE 6s WATCHES

Some of the most desirable (and expensive) American trench watches are the Giant size 6s. These watches are prized today both for their beauty and for their unusually large size. A WWI size 6s trench watch is about the size of a modern-day Rolex Datejust. Most American trench watches of the Great War were much smaller—size 3/0s or 0s. These two sizes ranged from 31mm up to 33mm in diameter, without counting the crown. A giant size 6s trench watch measures from 37mm to 40mm, making it highly sought after by collectors.

The three main manufacturers that provided watch cases for size 6s movements were The Illinois Watch Case Company (IWC Co), The Philadelphia Watch Case Company, and Bates & Bacon (B&B). The gold-filled Bates & Bacon size 6s trench watch case looks identical to the Philadelphia Watch Case Company silverode and sterling silver size 6s trench watch cases in size and dimension, and there's a reason for that: in 1901 Philadelphia bought Bates & Bacon. B&B was famous for high-quality solid gold and gold-filled cases. The Bates & Bacon version only came in 10K yellow gold fill; it was their Royal series. The Philadelphia cases were only manufactured in silverode and sterling silver. Silverode is a nickel alloy consisting of nickel, copper, manganese and possibly zinc. The Illinois size 6s trench watch cases were made of 10K gold-filled, solid nickel, and sterling silver. Of all these cases, the rarest one by far is the solid nickel Illinois case, which is semi-hermetic with swivel lugs.

Manufacturing numbers for the size 6s cases are not available, but I estimate there are about 80 percent fewer of these than the size 3/0s and 0s. You can expect to pay double or triple for a fully restored size 6s Elgin trench watch compared to the smaller size 3/0s and 0s simply due to their bigger size.

One thing to look out for when purchasing one of the 6s trench watches with a Philadelphia or Bates & Bacon case is the inner sleeve that holds the stem and crown in place. The inner sleeve gets screwed into the crown tube. This inner sleeve is very difficult to obtain these days, and it can make or break a watch if it has seen better days. If you are buying one in person, be sure to pull the crown in and out several times to ensure that the crown has a nice firm "pop" between the winding and setting positions. If it doesn't, and is very loose between these positions, I would advise not buying the watch or offering several hundred dollars less for the watch. This single part can be quite expensive and hard to find. If you are buying one from an online auction, be sure to ask the seller about this common issue before you place your bid.

Opposite page:
Top: Elgin "General Joffre"
Middle left: Threaded IWC Co size 6s case back
Middle right: The back of the dial is date-stamped.
Bottom: Take a look at the side view; the threaded sections of htis semi-hermetic case are nearly seamless. This example is nearly flawless.

Of the Giant size 6s Elgin trench watches, the best are the semi-hermetic cases with swivel lugs that were manufactured by the Illinois Watch Case Company. Semi-hermetic cases have a bezel and case back that are threaded, meaning that they screw on and off rather than pop on and off. This is a *far* superior design and was an early attempt to keep the elements out while in the field.

1918 ELGIN, GIANT SIZE 6s

The Illinois Watch Case Company manufactured this design in 1917 and 1918. This Elgin features a date-stamped dial proving its age; it was manufactured in June of 1918. The original center reservoir military hands have been relumed, as have the factory drilled hour marker lum dots. The enamel dial features a BOLD Arabic font. It has a grade 286 movement with 7 jewels. The lug diameter is 14mm wide, which was the standard for swivel lug cases manufactured by the IWC Co. Black pigskin strap with white stitching. The case measures a whopping 50mm lug to lug and 39mm without the crown. In September of 1917, the IWC Co named their size 6s cases with swivel lugs the "General Joffre."

Value: $1,700–$2,000
Rarity: 10+

1917 ELGIN, GIANT SIZE 6s

This trench watch with a 10K yellow gold-filled case made by Bates & Bacon (B&B). It is identical to the Philadelphia silverode size 6s case in size and shape. The Bates & Bacon cases were only available in 10K gold fill, and this Royal example is in pristine condition. B&B Royal grade cases were guaranteed for 20 years. The case measures 42mm lug to lug, 38mm without the crown with a 16mm lug diameter. The enamel BOLD Arabic military dial has a date-stamp on the back side of "11-17," or November 1917. The dial features factory-drilled hour marker lum dots. The original hands are also from 1917; the giveaway is the teardrop-shaped lum reservoir on the hour hand. The 7 jewel grade 286 movement is from 1917 as well. Very few of these 10K gold-filled B&B cases are in circulation; over the past years, I've only seen four of them.

Value: $1,500–$1,700
Rarity: 10+

10K Bates & Bacon yellow gold-filled Royal size 6s case.

Original 1917 military hands; note the teardrop shape of the center lum reservoir on the hour hand.

1917 ELGIN, GIANT 6s, BOLD ROMAN NUMERAL RED XII DIAL

Dial has a "USA" stamp. Thin spade hands, 38mm x 42mm Philadelphia silverode case, factory crown, 15 jewels, size 6s, 16mm buffalo strap.

Value: $1,200–1,500
Rarity: 10

1918 ELGIN, GIANT 6s, ARABIC RED 12 DIAL

Dial has a "USA" stamp. Thin spade hands, 38mm x 42mm Philadelphia silverode case, replacement crown, 15 jewels, size 6s, 16mm buffalo strap.

Value: $1,200–$1,500
Rarity: 10

1917 ELGIN, GIANT 6s, SHADOW BOX RED 12 DIAL

Dial has a "USA" stamp. Thick spade military hands, 38mm x 42mm Philadelphia silverode case, factory crown, 15 jewels, size 6s, 16mm buffalo strap. (slight chip on this dial, - $100)

Value: $1,200–$1,500
Rarity: 10

1917 ELGIN, GIANT 6s, BOLD ARABIC DIAL

Relumed skeleton hands and factory-drilled hour marker dots, 38mm x 42mm Philadelphia silverode case, factory crown, 7 jewels, size 6s, 16mm buffalo strap.

Value: $1,000–$1,100
Rarity: 9

1917 ELGIN, GIANT 6s, SHADOW BOX DIAL

Relumed center reservoir military hands (fat minute hand), 38mm x 42mm Philadelphia Silverode case, replacement crown, 7 jewels, size 6s, 16mm buffalo strap.

Value: $1,000–$1,100
Rarity: 9

1918 ELGIN, GIANT 6s, BOLD ARABIC DIAL

No second hand on dial, center reservoir military hands (fat minute hand), Illinois Watch Case Company solid nickel case with swivel lugs and hinged case back, factory crown, 7 jewels, 14mm pigskin strap.

Value: $1,000–$1,400
Rarity: 10+ (due to no second hand)

1918 ELGIN, GIANT 6s, SHADOW BOX DIAL

Relumed center reservoir military hands, 50mm x 37mm, 10K gold-filled Illinois Napoleon case with swivel lugs and hinged case back, factory crown, 7 jewels, 14mm pigskin strap.

Value: $1,650–$1,850
Rarity: 10+

1917 ELGIN, GIANT 6s, BOLD ARABIC METAL DIAL

Center reservoir military hands, factory-drilled hour marker lum dots, 38mm x 42mm Philadelphia silverode case, factory crown, 7 jewels, 16mm 1-piece nylon strap.

Value unrestored: $400 - $600
Rarity: 7

1915 ELGIN, GIANT 6s, BOLD ARABIC DIAL

Relumed center reservoir military hands and factory-drilled hour marker lum dots, 40mm x 47mm Philadelphia silverode case, factory crown, 15 jewels, 18mm strap.

Value: $1,000–$1,200
Rarity: 9

1917 ELGIN, GIANT 6s, SHADOW BOX DIAL

Reelumed center reservoir military hands, 37mm x 52mm Illinois Watch Case Company nickel case with swivel lugs and hinged case back, 7 jewels, 14mm pigskin strap.

Value: $1,200–$1,500
Rarity: 10

1917 ELGIN, GIANT 6s, BOLD ARABIC DIAL

Relumed skeleton hands and factory-drilled hour marker lum dots, 38mm x 50mm Illinois Watch Case Company sterling silver case with swivel lugs and a hinged case back, 7 jewels, 14mm pigskin strap.

Value: $1,650–$1,850
Rarity: 10+

1918 ELGIN, GIANT 6s, SHADOW BOX DIAL

Relumed center reservoir military hands, 38mm x 42mm, Bates & Bacon 10K gold-filled Royal case, replacement crown, 7 jewels, 16mm Buffalo strap.

Value: $1,500–$1,700
Rarity: 10+ (due to case)

Here are a few images of the watches that have had the lum replaced on the dial and on the hands with Bergeon Luminous Paste. In 1914, Elgin began applying luminous radium paint to their own dials in-house; the lum dots were drilled at the Elgin factory.

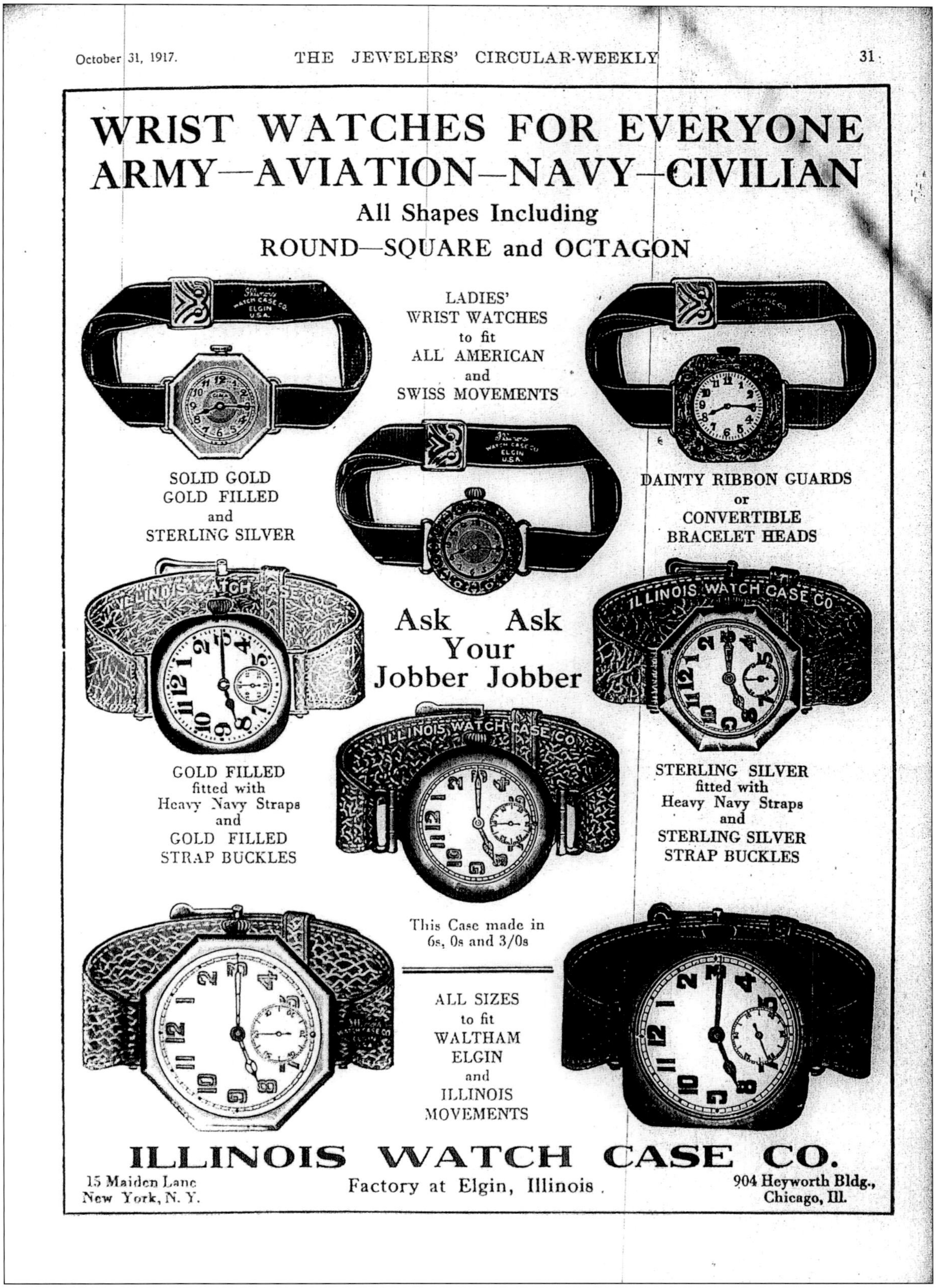

This ad for The Illinois Watch Case Company appeared in the October 31, 1917, issue of *The Jewelers Circular*. The bottom center illustration shows their beautiful size 6s military case design.

CHAPTER 6

THE RED 12 DIALS

The exact origins of the RED 12 dials have eluded watch collectors for years and will undoubtedly continue to do so for many years to come. Some say that putting a RED 12 on the dial was just an easy reference point for telling the time, some say that it was just for decorative purposes, while others will insist that it has military significance. I do not believe that anybody has found a definitive answer to this mystery. And once again, just like other Elgin military dials, the RED 12s were left out of the Elgin material catalogs for some strange reason. My best guess is that Elgin starting manufacturing size 0s RED 12 dials in 1912. I am basing this theory on several factors such as case design, age, movement, serial number, complete originality of the timepiece, and where the watch came from.

Elgin manufactured RED 12s with enamel dials as well as metal dials. I have cataloged no less than nineteen different RED 12 dial styles in enamel and metal. The RED 12s are very bold and beautiful, and they are highly sought after by collectors worldwide, so be prepared to pay a bit extra for one that is in pristine condition, without hairlines, chips, or cracks.

Not all RED 12 Elgin dials have a "Made in USA" or "USA" stamp on them; if they do, it will be just below the Elgin dial signature or just above the sunken sub-second dial. The dials that do have the "Made in USA" or "USA" stamps on them were probably manufactured for export markets such as Canada, New Zealand, Australia, England, Scotland, and Ireland.

RED 12 dials were very popular in Europe for many years after the Great War ended. Waltham was still manufacturing them well into the mid-1930s. Elgin probably stopped making them around 1925–1927, judging from examples that I have found around the world. To this day, if you walk into your local mall and look through the brand-new watch display cases, you will probably find a few RED 12s under the glass; they are still quite popular.

Six different Elgin RED 12 Dials

ELGIN RED 12 DIAL

The Elgin shown on this page is one of my favorites. Manufactured by the Illinois Watch Case Company, the sterling silver case has two rare features that make it stand above the rest. First, it is the larger size 0s, and second, it has swivel lugs. It is the only one of its kind that I have found. This style, known as the "cushion trench," is commonly found in the smaller size 3/0s but they have 14mm fixed lugs, not swivel lugs. This example still has the original sterling silver factory crown. The enamel RED 12 Arabic dial is also quite rare because it features a double "USA" stamp. The first "USA" is just below the Elgin signature and the second "USA" stamp is just below the 6 o'clock position. This Elgin has a brand-new olive drab webbing strap that I made from scratch and has been hand sewn. The original sterling silver hardware that was sewn into the old strap has been transferred to the new strap. The 1916 grade 354 Elgin movement features 15 jewels. In September, 1917, the IWC Co gave this case a proper name, the "Admiral Benson."

Value: $1,200 - $1,500
Rarity: 10 (due to case, strap and hardware)

Top left: February 6, 1918, ad
Top right: The Elgin "Admiral Benon" with swivel lugs.
Bottom right: Hand-sewn cotton webbing strap with original hardware.

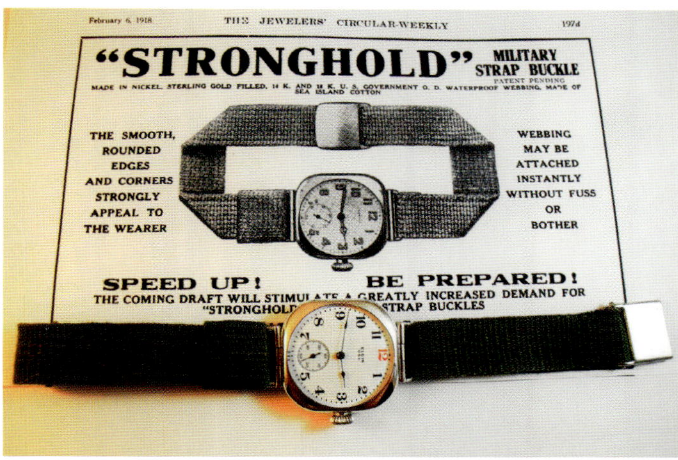

ELGIN, RED 12 BOLD ARABIC DIAL, ENAMEL

Of all the RED 12 Elgin dials, the one shown on this page is probably the most difficult to obtain. The numerals are about double the size compared to the smaller Arabic font RED 12s, which are much more common. I've only been able to locate a handful of watches with this dial over the years. The enamel dial has outer minute markers, a sunken sub-second hand, thick blue steel hands, BOLD Arabic numerals, and is signed ELGIN in all caps near the center with a RED 12 at the top. A "USA" stamp is just below the Elgin signature. It has a solid nickel, semi-hermetic case with a deep-well bezel that was manufactured by the Illinois Watch Case Company in Elgin, Illinois. This watch still has the original factory crown. 14mm tapered lugs with a genuine pigskin strap. 7 jewel movement from 1918, grade 462.

Elgin also manufactured the BOLD Arabic font RED 12 dial in metal; these are just as rare as their enamel counterparts.

Value: $900 - $1,000
Rarity: 9 (due to dial & case)

Elgin BOLD Arabic enamel RED 12 Dial with a "USA" stamp.

ELGIN, RED 12 BOLD ARABIC DIAL, METAL

The near-mint condition dial shown on this page is very rare. This dial has *not* been refinished; it's all original. A telltale sign that a metal dial has been refinished is the sub-dial where the second hand is located. The color of the sub-dial will have a bit of a shine to it if it is all original, while the rest of the dial will have a flat finish. This metal RED 12 dial was also available with a "MADE IN USA" stamp just above the second hand. This Elgin has the correct thick blue steel hands for this font style. The grade 462, 1918 7 jewel movement is set inside a sterling silver Illinois cushion trench case that still has the original sterling silver factory crown.

Value: $700 - $900
Rarity: 8 (due to dial)

BOLD Elgin Arabic metal RED 12 Dial

ELGIN, "ADMIRAL EVANS" RED 12

There are some trench watch case styles that collectors covet more than others due to rarity and style, and the 1918 Elgin shown here is among the most desirable. This is the Elgin "Admiral Evans" RED 12 Trench Watch. This case was manufactured by the Illinois Watch Case Company. It is a common misconception that this octagon case was exclusive to Illinois watch movements; it was also available for Elgin and Waltham movements. In September 1917, the IWC Co named this case the "Admiral Evans." The octagon cases were available with or without swivel lugs; the swivel lug versions are more desirable. These cases were available in sterling silver, gold-filled, and 14K solid gold; offset crown versions were also available. The pristine example shown here is made of sterling silver, and it still has the sterling silver original factory crown, which increases its value. The IWC Co octagon cases first came to market in mid-1917. I believe that these cases had a very short manufacturing run of only about two years. This size 3/0s Elgin has a grade 462 movement with 7 jewels that was manufactured in 1918. The enamel RED 12 dial features the BOLD Arabic font with a "USA" stamp just below the Elgin dial signature. It has the correct thick blue steel hands and second hand.

Value: $1,700–$2,000
Rarity: 10+

A 1918 Elgin "Admiral Evans" with swivel lugs. **Opposite: Middle left:** 7 jewel, grade 462 movement. **Bottom left:** Eight-sided sterling silver case. **Top right:** Note original factory crown.

ELGIN
U.S.A.

ELGIN
U.S.A.

21937965
ELGIN
U.S.A.

ILLINOIS
W.C.CO.
ELGIN
STERLING
5504836
30

ELGIN
U.S.A.

1915 ELGIN, RED XII

The Elgin trench watch on this page simply has the *best* of everything. This is a 1915 RED XII Elgin with a 14K solid gold case and 14K solid gold buckle with a very special 17 jewel movement. The 14K solid gold case with a hinged case back was made by the Illinois Watch Case Company early into the Great War. The case alone, without the movement or crystal, weighs in at 8.8 grams of 14K solid gold. As with most Illinois cases from this era, it has 14mm lugs. This grade 423, 17 jewel movement is incredibly rare. It is number 232 of the 500 made, serial number 18628232. Elgin manufactured about 2 million size 3/0s movements from 1914 to 1933; the vast majority of these had either 7 jewels or 15 jewels. Only 500 of the 2,000,000 were manufactured with 17 jewels. If you break those numbers down, only one-quarter of 1 percent were made with 17 jewels. The vertical demaskeening design on these 500 movements is unmistakable; the 7 jewel and 15 jewel versions looked nothing like this. Another feature that sets this movement apart is the material of the plates. They consist of solid nickel. The 7 jewel and 15 jewel versions were plated brass. Out of all of the size 3/0s movements produced, only grade 423 movements had a solid gold center wheel, solid gold raise jewel chatons, and solid gold screws on the balance wheel. The enamel dial on this Elgin features bold Roman numerals, an outer chapter of minute markers, a sunken sub-second hand, the correct blue steel hands, a bold Roman numeral RED XII, and a "MADE IN USA" stamp just above the second hand.

Value: $4,000–$5,000
Rarity: 10+

The 14K gold case was made by IWC Co. **Opposite: Top and middle:** Side views of the Illinois case and the crown; **Bottom:** Enamel Roman numeral RED XII Dial with "Made in USA" stamp.

ELGIN
MADE IN U.S.A.

1918 ELGIN, ARABIC RED 12 DIAL

Thin spade hands, Illinois Watch Case Company sterling silver case with an offset crown, size 3/0s, 7 jewels, 1-piece 14mm strap.

Value: $1250–$1,500
Rarity: 10+

1915 ELGIN, ARABIC RED 12 DIAL

Thin spade hands, Illinois Watch Case Company nickel case with swivel lugs, cushion-shaped bezel and round case back, factory crown, size 0s, 15 jewels, 16mm buffalo band.

Value: $800–$900
Rarity: 9

1917 ELGIN, SHADOW BOX RED 12 DIAL

Dial has a "MADE in USA" stamp. Enameled center reservoir military hands, "no-name" nickel case made by Wadsworth, factory crown, size 3/0s, 15 jewels, 3-piece 10mm bund strap.

Value: $750–$800
Rarity: 8

1915 ELGIN, BOLD ROMAN NUMERAL RED XII DIAL

Has a "MADE in USA" stamp, center reservoir military hands, Illinois Watch Case Company sterling silver semi-hermetic "General Funston" case with swivel lugs and an offset crown, factory crown, size 3/0s, 15 jewels, 1-piece nylon strap.

Rarity: $1,500–$1,700
Rarity: 10

1917 ELGIN, ARABIC RED 12 DIAL

Has a "MADE in USA" stamp, thin spade hands, Philadelphia silverode case, factory crown, size 3/0s, 7 jewels. Kitchener strap.

Value:$350–$400
Rarity: 4

1918 ELGIN BOLD ROMAN NUMERAL RED XII DIAL

Has a "USA" stamp, thin spade hands, Illinois Watch Case Company 10K gold-filled, semi-hermetic case with swivel lugs, size 0s 19 jewels, 14mm 3-piece bund strap.

Value: $900–$1,100
Rarity: 9

1916 ELGIN, SHADOW BOX RED 12 DIAL

Has a "USA" stamp, enameled center reservoir military hands (look at the minute hand), Wadsworth 14K gold-filled case with a hinged case back, factory crown, size 0s, 15 jewels, pigskin strap.

Value: $800–$1,000
Rarity: 9

1915 ELGIN, BOLD ROMAN NUMERAL RED XII DIAL

Thin spade hands, Bates & Bacon (B&B) Royal semi-hermetic 14K gold-filled case, factory crown, size 3/0s, 15 jewels, 12mm lugs.

Value: $700–$900
Rarity: 8

1918 ELGIN, BOLD ARABIC RED 12 DIAL

Has a "USA" stamp, thick hands, solid nickel case with swivel lugs made by the Illinois Watch Case Company, hinged case back, factory crown, size 3/0s, 7 jewels, 14mm pigskin strap.

Value: $800–$1,000
Rarity: 9

1917 ELGIN, ARABIC RED 12 DIAL

Thin spade hands, Dennison Watch Case Company nickel case with a hinged case back, factory crown, size 3/0s, 7 jewels, 16mm nylon strap.

Rarity: $500–$600
Rarity: 7

1913 ELGIN, ROMAN NUMERAL RED DIAL

Thin spade hands, Star Watch Case Company 14K gold-filled case with a hinged case back, replacement crown, size 0s, 7 jewels, 3-piece 10mm Bund Strap.

Value: $500–$600
Rarity: 7

1915 ELGIN, ARABIC RED 12 DIAL

Thin spade hands, Illinois Watch Case Company nickel case, cushion-shaped bezel and round case back, factory crown, size 0s, 15 jewels, 16mm nylon strap.

Value: $650–$750
Rarity: 8

1917 ELGIN, ARABIC RED 12 DIAL

Has a "MADE in USA" stamp, thin spade hands, Philadelphia sterling silver case with a hinged case back, replacement crown, size 3/0s, gilted movement, 7 jewels, 16mm buffalo band.

Value: $500–$575
Rarity: 6

1918 ELGIN, BOLD ROMAN NUMERAL RED XII DIAL

Thin spade hands, Illinois Watch Case Company nickel semi-hermetic case with swivel lugs and a deep-well bezel, factory crown, size 3/0s, 15 jewels, 14mm pigskin band.

Value: $1,000–$1,200
Rarity: 9

1917 ELGIN, SHADOW BOX RED 12 DIAL

Refinished metal dial, enameled center reservoir military hands, "no-name" nickel case by Wadsworth, factory crown, size 3/0s, 7 jewels, 3-piece 10mm bund strap.

Value: $400–$475
Rarity: 4

1916 ELGIN, ARABIC RED 12 DIAL

Thin spade hands, Bates & Bacon 14K gold-filled semi-hermetic case with a deep-well bezel, factory crown, size 3/0s, 15 jewels, 16mm lug diameter.

Value: $1,000–$1,200
Rarity: 9

1913 ELGIN ARABIC RED 12 DIAL

Double "USA" stamp, Moon hands, American Watch Case Company nickel silver case with an offset crown, factory crown, size 0s, 17 jewels, 12mm 1-piece strap.

Value: $575–$675
Rarity: 7

Same watch as above, with the bezel removed so the second "USA" stamp can be seen in the 6 o'clock position. I have seen about five Elgin dials over the years that have had a second "USA" on the dial.

1915 ELGIN, BOLD ROMAN NUMERAL RED XII DIAL

Has a "USA" stamp, thin spade hands, American Watch Case Company 14K gold-filled case (Canadian Division), hinged case back, factory crown, size 0s, 15 jewels, 12mm 1-piece strap.

Value: $650–$750
Rarity: 8

1917 ELGIN, ARABIC RED 12 DIAL

Has a "Made in USA" stamp, thin spade hands, Star Watch Case Company 14K gold filled case, hinged case back, factory crown, size 3/0s, 7 jewels, 12mm 1-piece strap.

Value: $550–$650
Rarity: 6

1917 ELGIN, SHADOW BOX RED 12 DIAL

Has a "USA" stamp (Swiss-made enamel dial), blue steel hands, Philadelphia silverode case with an offset crown, factory crown, size 0s, 17 jewel movement from 1911, 10mm bund strap.

Value: $750–$900
Rarity: 9

1910 ELGIN TRANSITION, ARABIC RED 12 DIAL

Has a double "USA" stamp, thin spade hands, Philadelphia silverode case, factory crown, size 0s, 7 jewels, 3-piece 10mm brown bund strap.

Value: $500–$600
Rarity: 6

1917 ELGIN, BOLD ROMAN NUMERAL RED XII DIAL

Has a "Made in USA" stamp, Illinois Watch Case Company sterling silver case, replacement crown, size 3/0s 15 jewels, 14mm pigskin band.

Value: $575–$675
Rarity: 7

1914 ELGIN, ROMAN NUMERAL RED XII DIAL

15 gram Keystone sterling silver case with a deep-well bezel, 1914 English market hallmark, 34mm case, factory crown, size 0s, 7 jewels, 10mm 3-piece black bund strap.

Value: $650–$850
Rarity: 8

1914 ELGIN, SHADOW BOX RED 12 DIAL

Has "Made in USA" stamp, center reservoir military hands, original lum on dial and hands, 10K gold-filled Wadsworth case with a hinged case back, factory crown, size 0s, 7 jewels, 16mm buffalo strap.

Value: $800–$900
Rarity: 9 (due to no second hand)

1913 ELGIN TRANSITION, BOLD ROMAN NUMERAL RED XII DIAL

Has "USA" stamp, thin spade hands, Philadelphia silverode case with an offset crown, factory crown, size 0s, 7 jewels, 12mm 1-piece strap.

Value: $650–$750
Rarity: 8

1918 ELGIN, BOLD ROMAN NUMERAL RED XII DIAL

Has "MADE IN USA" stamp, thick hands, Wadsworth sterling silver cushion case with spring bar lugs (case first available 1918), sterling silver factory crown, size 3/0s, 15 jewels, 16mm pigskin strap.

Value: $600–$750
Rarity: 7

1916 ELGIN, SHADOW BOX RED 12 DIAL

Has "USA" stamp, relumed skeleton hands, Philadelphia silverode case, factory crown, thick 16mm wide lugs, size 0s, 15 jewels, 16mm buffalo strap.

Value: $800–$1,000
Rarity: 9

1917 ELGIN, ARABIC RED 12 DIAL

Has a "Made in USA" stamp, Canadian market, thin spade hands, Empress 10K gold-filled American Watch Case Company Canadian Division (Toronto), hinged case back, factory crown, size 3/0s, 7 jewels, 10mm bund strap.

Value: $600–$700
Rarity: 7

ELGIN, BOLD ROMAN NUMERAL RED XII DIAL

Thick hands, "Empress" 10K gold-filled American Watch Case Company Canadian Division (Toronto), hinged case back, factory crown, size 3/0s, 7 jewels, 10mm bund strap.

Value: $600–$700
Rarity: 7

CHAPTER 7

THE BOLD ARABIC DIALS

One of my favorite Elgin trench watch dials is the BOLD Arabic enamel dial. Other than the RED 12 trench watch dials, these were considered among the first "military" style dials. Elgin only date-stamped dials on the back side that were considered "military" dials; the reason for this is unknown. Elgin began manufacturing these dials around 1913. The oversized BOLD numerals were very easy to read at a quick glance. These dials were available with a few variations, such as a BOLD RED 12 and factory-drilled hour marker lum dots. I even found an Elgin with all blue BOLD Arabic numerals, the only one that I have ever seen. Elgin made these dials in sizes 3/0s, 0s, and 6s for military wristwatch use. Here are a few pictures of the process of reluming the factory-drilled hour maker lum dots and the center reservoir military hands. The old radium lum has been removed from the dial and hands, and they have been properly cleaned.

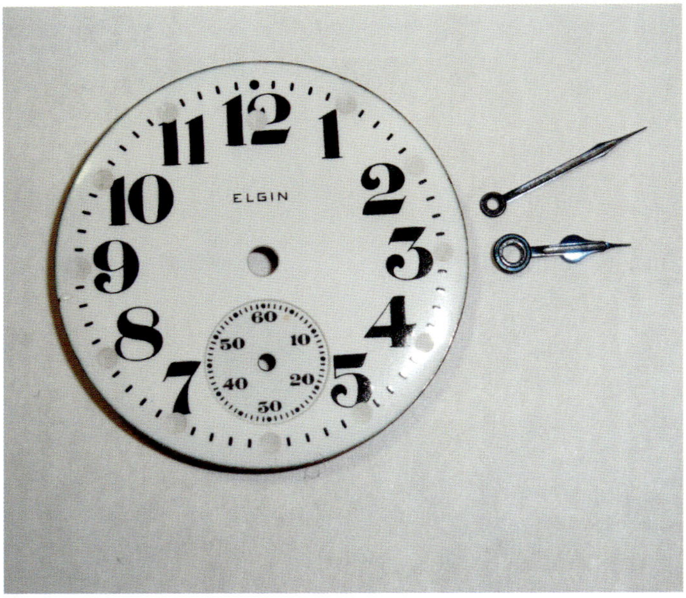

Enamel BOLD Arabic military dial with factory-drilled hour marker lum dots and center reservoir military hands.

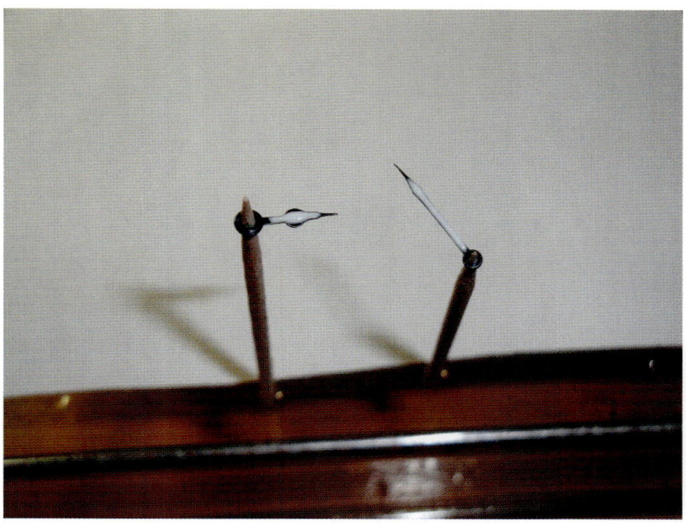

The new Bergeon luminous paste has been installed and is drying.

The new lum material is now dry and has been properly shaped for the hands. The watch has been fully reassembled and the crystal guard has been reinstalled.

ELGIN, BOLD ARABIC MILITARY NUMERAL DIAL

The beautiful example of an Elgin trench watch from the Great War (shown below and on the next page) is quite rare for several reasons. This is a solid nickel size 0s Illinois Watch Case Company semi-hermetic case with swivel lugs known as the "General Funston" case. These are quite hard to come by. The bezel and case back are threaded so they screw on and off, a far superior design. This is probably the largest size 0s case that you will find; it measures a whopping 50mm lug to lug and 34.5mm without the factory crown, 14mm lug diameter. The enamel BOLD Arabic military numeral dial is simply superb; it is the only BOLD Arabic dial within this book that has a "USA" stamp just below the Elgin dial signature. This is the only one of its kind that I have seen, and it is flawless. It still has the original thick blue steel hands and second hand. This Elgin also has the original factory crown. 14mm black genuine pigskin leather strap with white stitching.

Value: $700–$900
Rarity: 9

Elgin "General Funston," semi-hermetic case with swivel lugs.

This type of case was first available from the IWC Co in late 1917 just after the paperwork was filed with the U.S. Patent Office; they were manufactured with a "patent pending." This particular case was made after September 3, 1918, when the patent had been granted, and the inside of the case back is stamped with this information. This watch has a beautiful stance on the wrist—nice and large compared to other size 0s watches of the era.

The threaded case back.

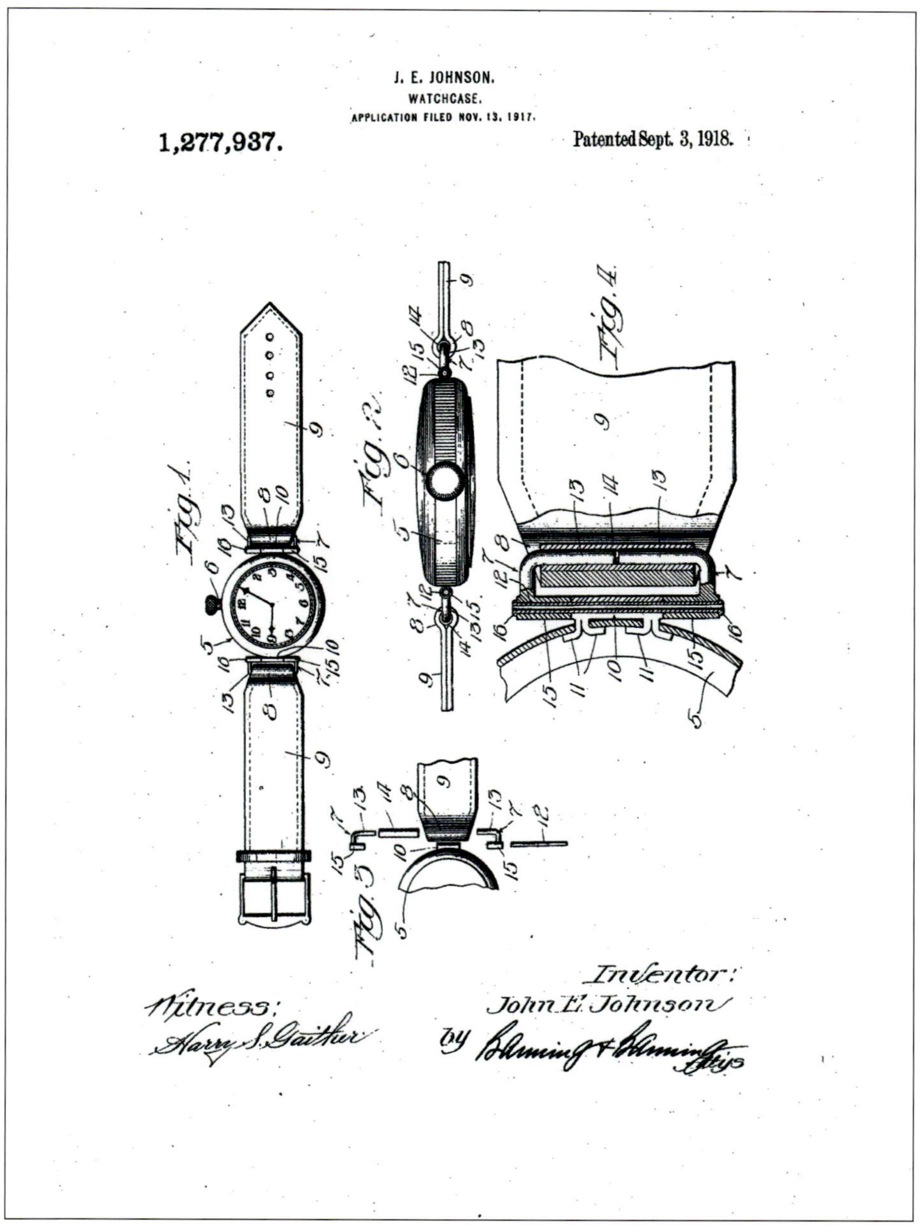

This watch case design was invented by John E. Johnson, an employee of the Illinois Watch Case Company in Elgin, Illinois. The patent application was filed on November 13, 1917, and the patent was granted on September 3, 1918. This is the drawing that was filed with the United States Patent Office for the wristwatch case. *Courtesy United States Patent Office.*

ELGIN, BOLD ARABIC ENAMEL 24-HOUR MILITARY DIAL

The watch shown below is quite special. It features a swivel lug case that was manufactured by the Philadelphia Watch Case Company and is very rare. Philadelphia cases almost always had solid fixed lugs, not swivel lugs, making this example the exception to the rule. The most prominent American manufacturer of swivel lug cases during the Great War was the Illinois Watch Case Company. This Philadelphia case is made from a silverode material that is incredibly durable. The case features 14mm swivel lugs and an oversized factory crown that is about 50 percent larger than your normal crown of this era. Having the original factory crown on this watch can make or break its value. The case measures 40mm lug to lug and 31.5mm without the crown. It has a 1918 grade 462 movement with 7 jewels, size 3/0s, thick blue steel hands and a 14mm pigskin strap.

Value: $1,250–$1,500
Rarity: 10+

An enamel 24-hour military dial.

Note the oversized original factory crown and the case model "9008" with swivel lugs.

ELGIN, BOLD ARABIC ENAMEL MILITARY DIAL

The Philadelphia Watch Case Company came out with several radically different case designs during the Great War; this design is one of them. This is a silverode case that is huge for a size 0s. It measures 45mm lug to lug and 34.5mm without the crown, 19mm lug diameter. This watch still has the original oversized factory crown, which greatly impact the watch's value. Philadelphia made this case design in size 0s and the even larger size 6s. These cases are quite hard to come by; I have only seen one in a size 0s and three in size 6s. The piece has a beautiful stance on the wrist due to its larger size. It features a grade 320 movement with 7 jewels from 1914, a BOLD Arabic enamel military dial, Moon hands, and the original second hand.

Value: $900–$1,000
Rarity: 10

This watch has a very wide, 19mm lug diameter and a case that is 34.5mm wide. It also has the oversized original factory crown.

ELGIN, BOLD ARABIC ENAMEL DIAL IN GOLD-FILLED CASE

In some instances the watch case design is what makes the watch extremely valuable. Sometimes a gold-filled watch case can be more valuable than another design that is made of solid gold. The size 0s "Giant" 10K yellow gold-filled barrel case made by the Illinois Watch Case Company (shown here and on opposite page) is a perfect example. The barrel cases from Illinois are quite hard to find, and this one does not have any gold fill wear spots. The sterling silver versions weigh about 21 grams; this gold-filled version weighs about 20 grams without the movement. This case is rather large, measuring 42mm lug to lug, 33mm without the original factory gold-filled crown with a 20mm lug diameter. This Elgin features a 1916 grade 320, size 0s movement with 7 jewels. The BOLD Arabic enamel dial has factory-drilled hour marker lum dots that have been refilled with Bergeon luminous paste, as have the original skeleton hands.

Value: $1,500–$1,800
Rarity: 10

Size 0s 10K gold-filled IWC Co "Giant" barrel case.

Size 0s grade 320 movement with 7 jewels.

1914 ELGIN, BOLD ARABIC DIAL

Moon hands, no drilled hour markers, Philadelphia Silverode case, factory crown, size 0s, 7 jewels, 1-piece 12mm leather strap. U.S. Army Cavalry watch.

Value: $700–$900
Rarity: 8

1914 ELGIN, BOLD ARABIC DIAL

Skeleton lumed hands and factory-drilled hour marker dots, Fahys "Montauk" 10K gold-filled case, factory crown, size 0s, gilded movement, 7 jewels, 12mm pigskin strap.

Value: $850–$950
Rarity: 9

1916 ELGIN, BOLD ARABIC DIAL

Lumed center reservoir military hands and factory-drilled hour marker dots, Illinois nickel silver case with a hinged case back, nonfactory crown, size 0s, gilded movement, 7 jewels, 12mm pigskin strap.

Value: $400–$500
Rarity: 5

1916 ELGIN, BLUE BOLD ARABIC DIAL

Thick spade hands, no drilled hour markers, Philadelphia silverode case, factory crown, size 3/0s, 7 jewels, 3-piece 10mm bund strap.

Value: $1,300–$1,500
Rarity: 10+ (due to blue numerals)

1922 ELGIN, BOLD ARABIC ENAMEL DIAL WITH RED FACTORY 24-HOUR MARKERS

Made post WWI, but also available during the war. Thick spade hands, 9K solid gold Keystone Watch Case Company Case with a hinged case back and a coin-edge bezel, size 3/0s, 7 jewels, 3-piece 10mm bund strap.

Value: $1,400–$1,600
Rarity: 10

1915 ELGIN, BOLD ARABIC DIAL

Skeleton lumed hands and factory drilled hour marker dots, Illinois nickel case with a hinged case back, factory crown, size 0s, 15 jewels, 3-piece 10mm bund strap.

Value: $400–$500
Rarity: 6

1918 ELGIN, BOLD ARABIC DIAL

Thick spade hands, no drilled hour marker dots, Philadelphia "Silverode" case with an offset crown, factory crown, size 3/0s, 7 jewels, 3-piece 10mm bund strap.

Value: $500–$600
Rarity: 7

1917 ELGIN, BOLD ARABIC DIAL

Thick spade hands, no drilled hour marker dots, Illinois sterling silver case, factory crown, size 3/0s, 7 jewels, 14mm pigskin band.

Value: $500–$600
Rarity: 7

1917 ELGIN, BOLD ARABIC ENAMEL DIAL WITH RED FACTORY 24-HOUR MARKERS

Thin spade hands, Illinois sterling silver case, factory crown, size 3/0s, 7 jewels, 14mm pigskin band.

Value: $750–$900
Rarity: 8

1915 ELGIN, BOLD ARABIC DIAL

Thick spade hands, drilled hour marker dots, Philadelphia silverode case with an offset crown, factory crown, size 0s, 15 jewels, 3-piece 10mm bund strap.

Value: $400–$500
Rarity: 6

1913 ELGIN, BOLD ARABIC DIAL

Moon hands, drilled hour marker dots, Wadsworth Watch Case Company 14K gold-filled case with a hinged case back, factory crown, size 0s, 17 jewels, 3-piece 10mm bund strap.

Value: $450–$550
Rarity: 6

1914 ELGIN, BOLD ARABIC DIAL

Thick spade hands, Philadelphia Watch Case Company silverode case, replacement crown, size 0s, 7 jewels, 12mm pigskin band. U.S. Army Cavalry watch.

Value: $450–$550
Rarity: 6

1916 ELGIN, BOLD ARABIC DIAL

Thick spade hands, Wadsworth sterling silver case with an offset crown, 16mm lugs, factory crown, size 0s, 7 jewels, gilded movement, 16mm leather strap.

Value: $600–$700
Rarity: 7

1917 ELGIN, BOLD ARABIC DIAL

Lumed center reservoir military hands, factory-drilled hour marker lum dots, 34.5mm semi-hermetic solid nickel Illinois case with a coin-edge bezel, factory crown, size 0s, 7 jewels, 12mm pigskin strap.

Value: $750–$850
Rarity: 9

1917 ELGIN, BOLD ARABIC DIAL

Thick spade hands, fat lug Philadelphia silverode case, 16mm lugs, factory crown, size 0s movement that pre-dates the case by three years, 7 jewels, 16mm leather strap.

Value: $500–$600
Rarity: 7

RELUMED WATCHES

Here are a few images of watches that show how the lum has been replaced on the dial and on the hands with Bergeon luminous paste.

The two Elgin Wrist Watch advertisements shown here, featuring BOLD Arabic Dials for Outdoor Men, are very special as they are the earliest ones that are known to exist. The ad on this page was published in *The Keystone* magazine on September 1, 1913, and it features Louis Disbrow. Disbrow raced in the first four Indianapolis 500s. Elgin was light-years ahead of its time in using famous race car drivers to promote its watches. This type of marketing campaign is still very popular today.

The ad on the opposite page was published October 29, 1913, in *The Jewelers' Circular*, and it features race car driver "Wild Bill" Endicott in his "Case Tornado." Wild Bill took fifth place in the 1912 Indianapolis 500. He is shown sporting his 1913 Elgin wristwatch that has a BOLD Arabic dial without a second hand. Disbrow and Endicott competed in the annual Elgin National Road Races for the Elgin National Watch Company trophy.

A September 1, 1913, ad in *The Keystone*.

An ad from October 1913 in *The Jewelers' Circular.* Weekly.

After searching through hundreds of weekly and monthly publications from the Great War period, I found the Elgin advertisement below in the June 13, 1914, edition of the *Saturday Evening Post,* page 31. A nearly identical ad was in the June 3, 1914 edition of *Life* magazine, but the copy I saw was of poor quality. Wristwatches for men at this time were not very popular; the general consensus was that only a woman would wear a wristwatch. Over the next year, Elgin launched a massive campaign to change that idea, stressing that a "real men" wore Elgin wristwatches. Notice that the June 13, 1914, ad takes up just one-quarter of a page in a major publication. In later months, Elgin took out full-page ads in *Outdoor* magazine.

The next two pages show five Elgin wristwatch advertisements that were featured in *Outdoor* magazine in 1915. The May 1915 and October 1915 ads state that the Elgin watch with the BOLD Arabic enamel dials and Philadelphia cases were purchased and worn by men in the U.S. Army Cavalry Division during the Great War.

THE SATURDAY EVENING POST · 31

MEN'S WRIST WATCH
Pigskin strap. Nickel, $10; silver, $12; gold filled, $15.

Elgin WRIST WATCHES

Necessary equipment for the outdoor man and woman. Accurate, handy, beautiful timekeepers, always in plain sight and cannot be lost. Strenuous activity will not affect them and dampness cannot harm them.

By all means get one for your vacation this summer and save your expensive watch from possible damage or loss. See them at your Elgineer's store—your local jeweler's.

ELGIN NATIONAL WATCH CO.
ELGIN, ILLINOIS

WOMEN'S BRACELET WATCH
$20 to $75.

grinningly over the breakfast table the next morning to his mother. And there would have been no significance attached by any one—except perhaps Cora's mother—had not Lil Todd a week later been hard pressed to fill out the five pages of correspondence that friendship with a married schoolmate in Kansas City exacted from her monthly.

Like most talkative people, Lil loathed letter writing. She groaned at the labor of remembering interesting items and jotted them down in hodgepodge connection. And so, merely to start a page, she mentioned that Rudy had Corry Kaley as partner in the grand march of the midwinter dance.

Jenny, the married schoolmate, remembered Cora as a pretty, self-contained girl—not as discard. She knew that Rudy Warner always led the grand march. She interpreted the news as significant, and passed it on with her interpretation. And so in less than two months, through the mouths and pens of five or six more or less interpretative people, it finally reached Kel Holmsted, who had wandered far and wide without locating the prospects that Jannsville had not offered.

Kel was surprised. He had always pictured Corry as patiently teaching school and waiting until he should return—which he had no intention of ever doing. Why, Corry was old! She had been old when he left! Well! Well!

Now it happened that when Kel heard the news he had been homesick for some time—if a lazy, calfish, forlorn feeling is entitled to insult that decent, sad word. And for some time he had rather resented the fact that Cora Kaley, so to speak, kept him from going home. He had wished that she would move away so he might return. He took the news of her engagement to Rudy as a glad removal of a barrier.

And—by jingo!—Corry had done well for herself! Kel had always trailed along in the pale orbit of conquest that lay outside Rudy's glowing center; and, despite a very fair opinion of his own desirability, he knew that, matrimonially considered, Mr. Warner outclassed him as Jannsville's one taxicab outwheeled its hack. Corry was a lucky girl, by George! He would congratulate her! He wondered, though, what Rudy saw in her. She had been sort of pretty years ago, before she got thin and oldish; but no style—not a mite! Maybe, though, she had perked up. Well, anyway, he would go back; and he bought a ticket and started.

Helena, Montana, where Kel happened to be when he bought his ticket, is a long journey from Jannsville. So many days on a train gives you time for thought—doubled-over, crisscross, labyrinthic thought. Kel's mind traveled the usual barking, circling, winding paths pursued by a certain grade of masculine mind when it learns that the fruit it threw away as specked or vapid has been picked up by a connoisseur and adjudged of rare delicacy.

The first day, Kel framed congratulations to Corry. The second, he wondered, with mirth, how she had landed Rudy. The third, he wondered—without mirth. The fourth, his mouth curled cynically. And all this time he had been under a delusion that she cared for himself! He had even felt sorry for her. Huh! Guess no man need waste pity on a woman! What fickle things women are!

On the fifth day Kel Holmsted wrapped himself in aggrievement, like an abandoned orphan. And this was the tenor of his thought: Oh, faithfulness! How art thou departed from this earth!

The sixth—why, certainly! He was the injured party—had been all these years! She wrote only two letters in reply to that last postcard, and a girl who thought anything of a man would have feared that those two might have miscarried and she would have written to find out whether he got them. What good coconut-cream pie her mother used to make every Sunday! Kel's mouth watered. Montana—the part he knew—was as shy of good pie as of coconut trees.

On the seventh day, however, Kel nobly and sadly decided not to interfere—unless his feelings overmastered him at sight of the guilty pair. Under the nobility of resolution lay a fear that his feelings would be unmanageable.

Meantime Rudolph had not sold the Kaley cottage. He was out of town a week or so, at home with a cold a week, out of town again for a few days. On his return he found a curt note. What had he done?

It appeals to the most critical taste

There is no guest too fastidious, no occasion too formal or too important for

Campbell's Tomato Soup

Its irreproachable quality commends it to the most exacting epicure. And it is so readily adapted to the character of the meal—light or hearty as the case may be—that in each instance it seems exactly the fitting key to the occasion.

Prepared as a rich Cream-of-tomato, or served in bouillon cups topped with whipped cream it is an especially inviting "overture" to any meal.

How about that particular "company-affair" of yours?

21 kinds 10c a can

Campbell's Soups
LOOK FOR THE RED-AND-WHITE LABEL

The Saturday Evening Post, June 13, 1914.

May 1915 , *Outdoor* magazine.

October 1915, *Outdoor* magazine.

September 1915, *Outdoor* magazine.

November 1915, *Outdoor* magazine.

December 1915

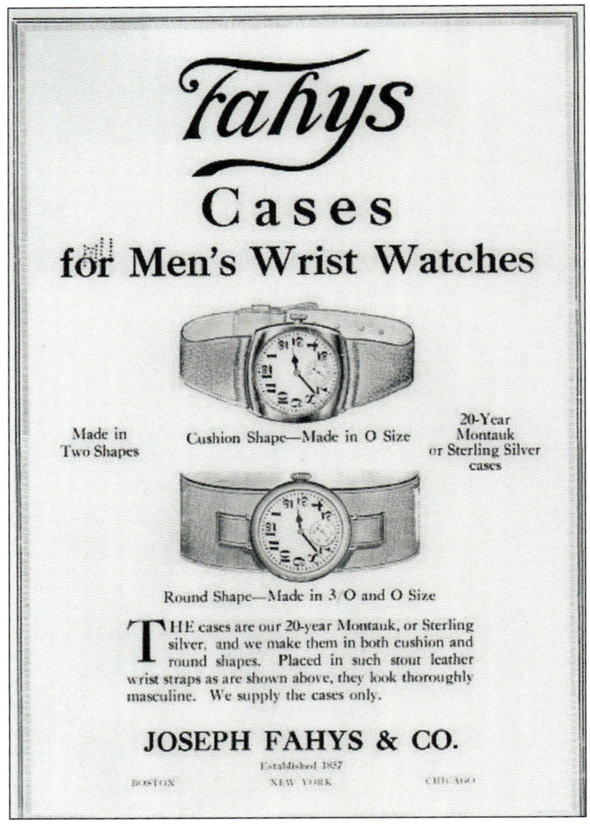

These cases from Joseph Fahys & Co. feature the BOLD Arabic Dials.

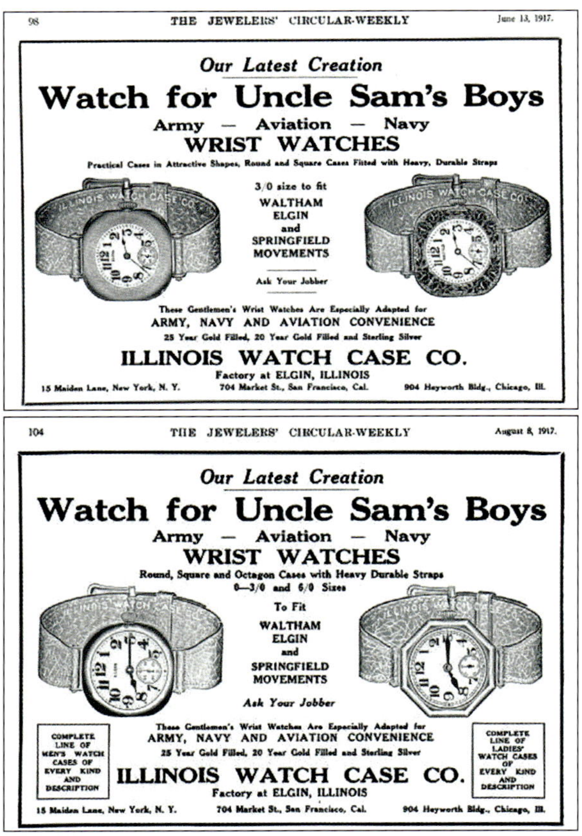

These ads from the Illinois Watch Case Company appeared in the June 13, 1917, and August 8, 1917, issues of *The Jewelers' Circular*. Both feature BOLD Arabic dials.

The Star Watch Case Company also liked to use the BOLD Arabic Dials in their ads. This one appeared in the January 30, 1918 edition of *The Jewelers. Circular*.

CHAPTER 8

SHADOW BOX DIALS

The black and white Elgin Shadow Box enamel military dials are probably the most recognizable of all of the Elgin trench watches from the Great War, aside from the Star Dial trench watches. Elgin manufactured more of these than of any other military dial.

Production on these dials began in 1916 and continued until around 1924. The black-and-white shadow box military dials came in two different styles, the only real difference being the size of the numeral font. Both had an outer chapter of minute markers and a sunken sub-second dial. The font on the larger numeral dial is just a bit bolder as well.

This Elgin Shadow Box Dial once belonged to Lt. Philip Niederaur, who honorably served in the U.S. Navy during the Great War.

Shown here, side by side, are the two Shadow Box font sizes.

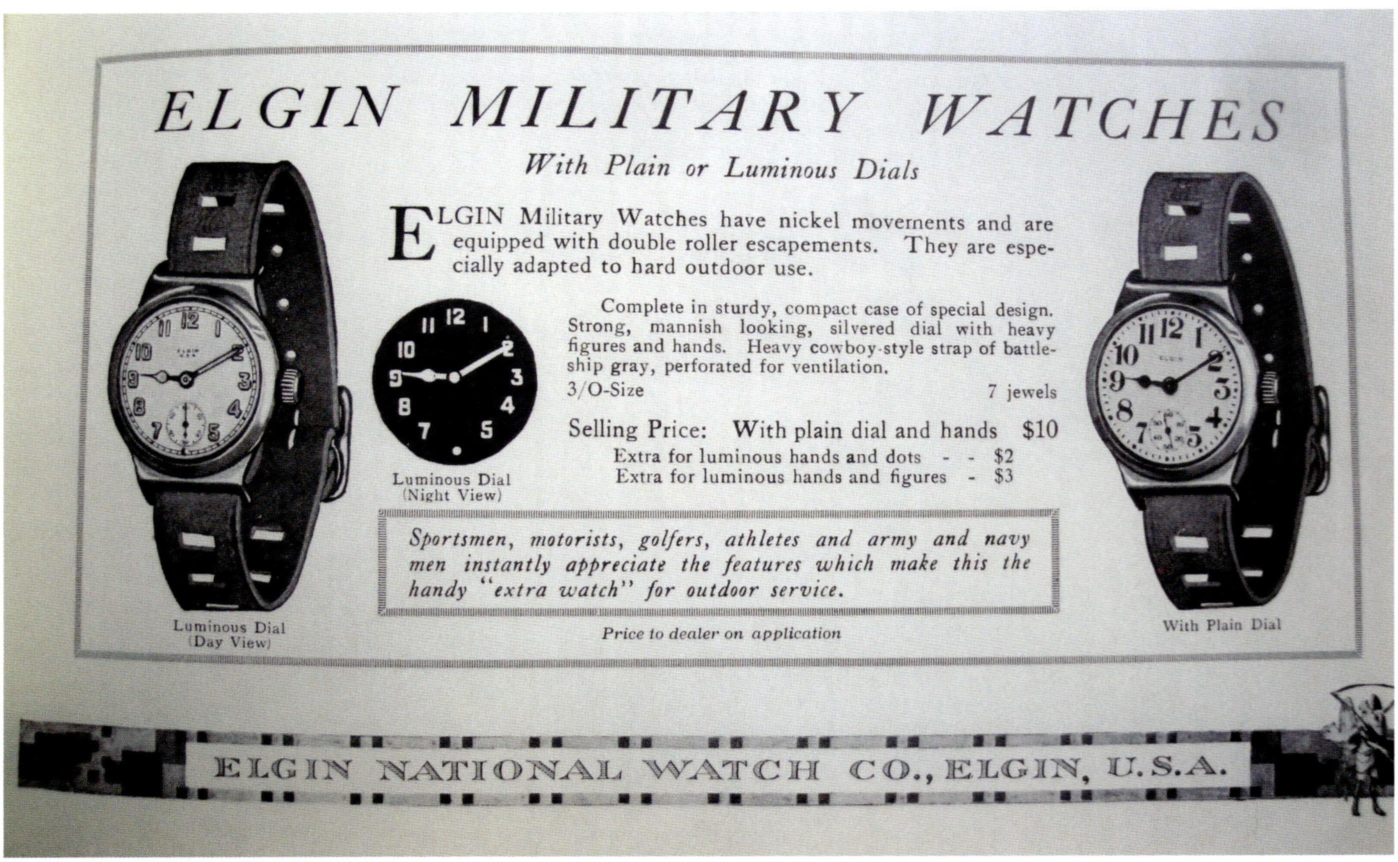

This ad for The Elgin National Watch Company that appeared in the 1917 Benjamin Allen & Company catalog (a wholesaler in Chicago). It shows two beautiful Elgin military watches. The one on the left has a Shadow Box Dial and the one on the right has the BOLD Arabic Dial. What makes this ad quite rare is that the dial on the left has a "USA" stamp just below the Elgin dial signature. It's the only Elgin ad that I have seen that has a "USA" stamped dial.

The Elgin pictured in the ad at right was number 19 on *Outing* magazine's best Christmas gifts for the year 1916. It was featured in the December 1916 issue. The ad states that the U.S. Army adopted the Elgin model shown for the infantry and cavalry divisions. It could be purchased for $13 with luminous hands and dial. This would be a big change from the watches being worn by the U.S. Army Cavalry divisions from 1914 to 1916. In those years, they were wearing the Elgins with the BOLD Arabic enamel dials. The 1914 BOLD Arabic Elgins were a size 0s. The watch pictured here is the smaller size 3/0s. This watch had one of those "no-name" cases that was just stamped "nickel" with a serial number. I have concluded that the Wadsworth Watch Case Company was in fact the manufacturer of the "no-name" nickel cases. I find these cases to be of superior design and quality. They snap together very crisply and are rather hard to open unless you know what you are doing.

A Night and Day Military Watch (19)

THE sportsman of to-day who carries his watch in a pocket is in the same class as the man who makes use of a flint and steel to light his pipe. To say that he is behind the times is to underestimate the case.

My own experience justifies the belief that the out door man's hands are occupied pretty much all of the time—so much so at any rate,—as to make the process of fishing for his watch a downright nuisance. Especially is this true in the winter when telling the time is often a matter of removing one's mittens and burrowing under layers of clothing.

No matter what the season, however, the only logical place for the sportsman's time-piece is where it can be consulted instantly under any circumstances—hence the wrist-watch. If common sense itself did not suffice to establish the wrist as the most practical place to wear a watch the testimony of hundreds of thousands of men on the battle fields of Europe would do so beyond question. Uncle Sam has recognized this truth and in doing so has adopted the model shown herewith—the Elgin Military Watch—for use in both infantry and cavalry.

Fitted with a heavy balance and double-roller escapement (the better to withstand rough usage), the works are enclosed in a sturdy case, dull finished to prevent reflected light from betraying the wearer's whereabouts. The watch is fastened to the wrist with a strong leather strap of battle ship gray and is unusually good to look at.

The Elgin may be read as easily on the blackest night as at noon for hands and figures are luminous and stand out clear and clean-cut in inky darkness—which feature is of inestimable value on the trail and a great comfort at all times.

I know of no gift that will meet with greater appreciation from true sportsmen. As described above the watch costs $13.00. Without the luminous feature, it retails for $10.00.

Catalogs, such as this one from Benjamin Allen & Company from 1917, are superb ways to see exactly what the watches looked like when they were brand new.

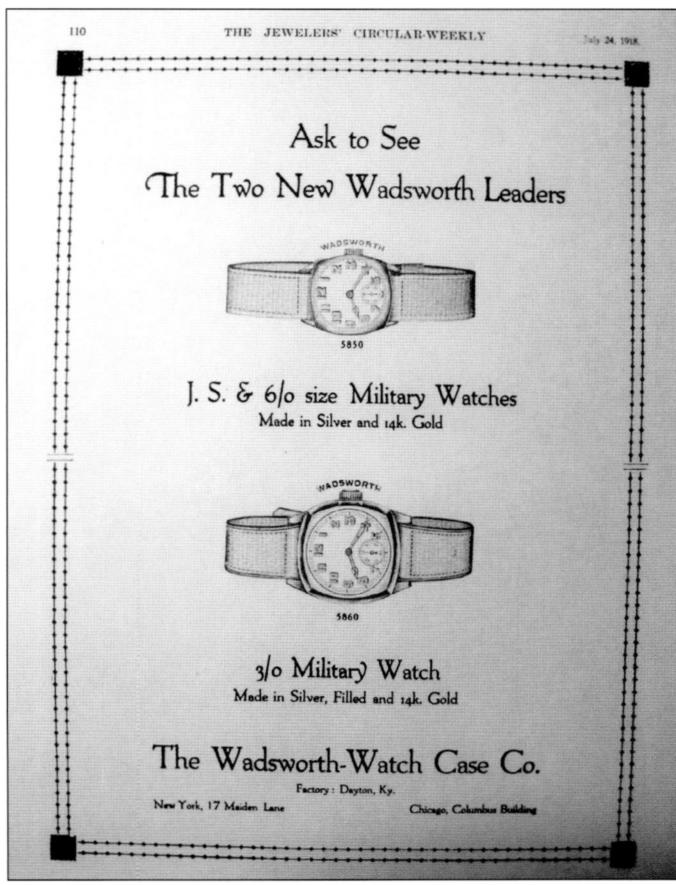

Watch case styles were starting to change toward the end of The Great War, but the dial styles remained the same for a few more years. This ad for the Wadsworth Watch Case Company shows a cushion case wristwatch with spring bar lugs and a lumed shadow box dial from July 24, 1918. It is a common misconception that this style of case was not available until 1922 or 1923. Wadsworth was way ahead of its time when it came to new case design and style.

Far left:

This advertisement from the Keystone Watch Case Company dated September 1918 is one of the rarest known. Keystone did very little advertising for their watch cases during the Great War. The Philadelphia Watch Case Company was owned by the Keystone Watch Case Company. All four of these cases could have been stamped Philadelphia or Keystone on the inside of the case back. They only manufactured one case style during the war with swivel lugs, and it is on the far right. Finding an ad featuring an offset crown during this era is almost impossible; this ad has two of them, making it very special. Both of the offset crown cases have a coin-edge bezel.

Left:

This detail from the Keystone Watch Sept. 1918 ad is the only case that I can think of that was first produced with solid fixed lugs then crossed over to new technology and started using a spring bar lug design. Spring bar lugs began gaining popularity in mid-1918. Wadsworth, Philadelphia, Star, and Gruen all had spring bar lug case designs available when the Armistice was signed in November of 1918.

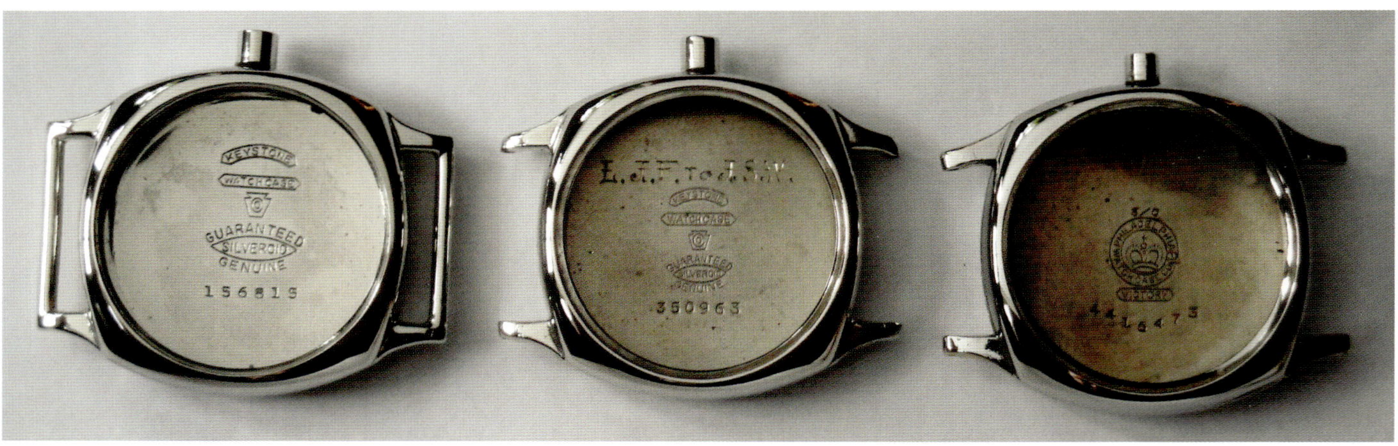

The image above shows the evolution of case number 9006 starting out with solid fixed lugs then crossing over to spring bar lugs. The case materials pictured are silveroid and 10K white gold filled (Victory). Silveroid is a nickel alloy. Note that the gold-filled case on the far right has a Philadelphia stamp, not a Keystone stamp.

In later years, this case design also had an added feature: engraving on the sides of the case. This case was manufactured circa 1919-1923.

This ad from M. Sickles & Sons, dated March of 1918, features one of the most elusive cases manufactured by the Illinois Watch Case Company, the offset crown cushion case, which they began making in 1917. This is only the second ad that I've seen that features a case with an offset crown, and the only cushion case design with an Elgin movement and shadow box dial depicted.

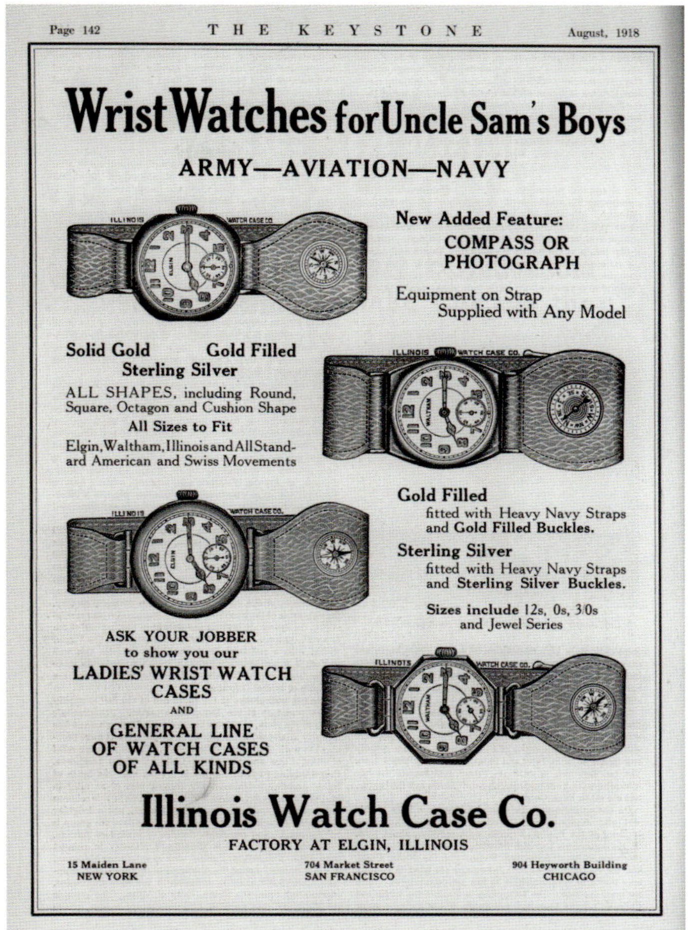

The Illinois Watch Case Company mainly used BOLD Arabic Dials and Shadow Box Dials in their advertisements. This ad in *The Keystone* features Shadow Box Dials. Four different case shapes are represented: cushion, barrel, round, and octagonal. All were available in solid gold, gold-filled, and sterling silver.

This advertisement from the Illinois Watch Case Company is quite important as it shows the octagonal case on the bottom right with lugs that have a pin. The ad above showed the octagonal case with swivel lugs, the far more rare version.

ELGIN, SHADOW BOX DIAL WITH STERLING SILVER CASE

The Illinois Watch Case Company had some case designs that simply outweighed the others. An average American-made size 0s WWI sterling silver trench watch case weighs around 9 to 11 grams depending on the manufacturer. The sterling silver Illinois case, shown on this page and the next page, weighs a whopping 21 grams without the movement, dial, hands, crystal, and strap. It's a very elegant barrel design, and the case still has the original factory crown. The case measures 42.5mm lug to lug, 33.5mm without the crown with a 20mm lug diameter. The original center reservoir military hands have been relumed. This Elgin features a grade 320 transition movement with 7 jewels, but the Illinois case was manufactured in 1917. This case style was available in sizes 3/0s and 0s. These cases are quite rare.

Value: $1,100–$1,300
Rarity: 10+

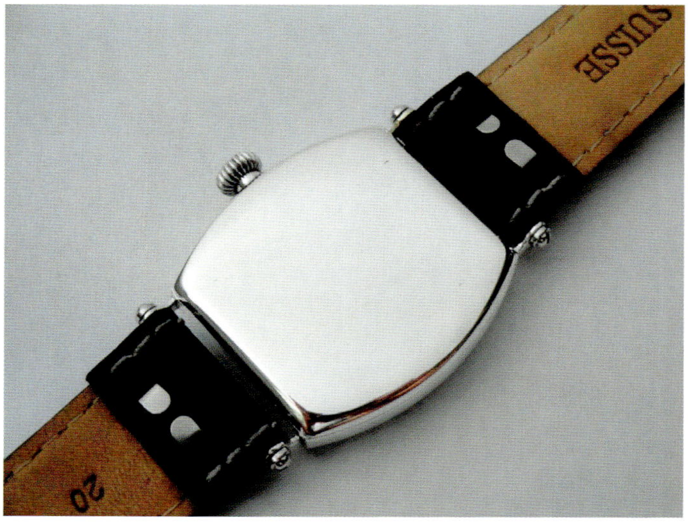

This sterling silver barrel case made by the Illinois Watch Case Company weighs a whopping 21 grams, twice as much as similar sterling cases for American trench watches.

These close-ups show the relumed original military hands and the original factory sterling silver crown.

ELGIN, SHADOW BOX DIAL MILITARY DIAL

The Star Watch Case Company was originally located in Elgin, Illinois. In 1897, they began manufacturing pocket watch cases, then in 1905 the company moved to Ludington, Michigan. Star was mainly known for manufacturing round gold-filled wristwatch cases during the Great War. They also had a few other designs, one of them being quite special. The pristine example pictured here is their cushion trench watch case design. What sets this design apart from the Illinois Watch Case Company or Joseph Fahys cushion trench design are the lugs. The Star lugs go all the way out to the ends of the case, and they are very thick at 2.25mm. This case has a very wide 16mm lug diameter, while the other manufacturers' versions had 14mm lugs. This case was available in sterling silver and in gold-filled. The case measures 40mm lug tip to lug tip and 31mm without the original factory crown. Size 3/0s, 15 jewels, enamel shadow box military dial with center reservoir military hands, green khaki one-piece strap. This exact case design with the wide lugs from Star is quite difficult to locate.

Value: \$700–\$800
Rarity: 8

The Star Watch Case Company made this sterling silver case with 16mm wide lugs.

When the United States entered the Great War in April of 1917, there was a massive need for wristwatches, but supply could not meet the demand. Companies like Elgin, Waltham, and Illinois were manufacturing movements as fast as they could. In late 1917, the Becker-Heckman company in Chicago, Illinois, stepped up to assist with the shortage problem by militarizing civilian stock. They re-cased older pocket watch movements into wristwatch cases and some pocket watch cases were converted into wristwatch cases. Becker-Heckman advertised nationwide in jewelry trade magazines that they could convert a retailer's nonsalable regular stock into military watches.

While it's not known whether Becker-Heckman did any work on the Elgin shown here, the watch is a perfect example of this common practice. The 7 jewel size 0s movement in this Elgin dates back to 1895, but the 10K gold-filled "Royal" wristwatch case that was made by Bates & Bacon is from 1917 or 1918. The enamel Shadow Box Dial on this Elgin movement was not available in 1895; it would not come out for another twenty years. This Elgin has a grade 109 "model 1" movement, but the dial was originally designed for "model 2" movements that were drastically different. The three dial feet hole locations were different on a "model 1" movement than on a "model 2" movement. In order for older "model 1" movements to be militarized with a Shadow Box Dial, the Elgin dial department had to move the dial feet locations. This is why you will see many Elgin trench watches with movements that pre-date the cases.

The Elgin movement in this watch dates back to 1895 but the 10K gold "Royal" case by Bates & Bacon is from 1917 or 1918.

1895 "model 1" Elgin movement

1916 "model 2" Elgin movement

This photograph shows the difference between "model 1" and "model 2" dial feet locations. "Model 1," on the left, has the dial feet locations at the 13 - 42.5 - 55 minute markers. "Model 2," on the right, has the dial feet locations at the 11 - 38 - 55.5 minute markers. This is why Elgin had to manufacture special dials for older "model 1" movements in order for them to be militarized.

This is an advertisement for the Becker-Heckman Company from December 5, 1917. Several variations of this ad exist. Many of them show different case style possibilities such as cushion, round, and barrel. A Wadsworth case is shown here..

ELGIN, LARGE FONT SHADOW BOX ENAMEL DIAL

Far fewer of the large font dials, like the one shown here, were manufactured than the small font. The 10K yellow gold-filled Bates & Bacon case features the very desirable 16mm wide lug configuration; the lugs are 2.25mm thick. It has a hinged case back and the original oversized factory crown. This case was made as a wristwatch case; it is not a converted pocket watch case. B&B manufactured this case in sizes 3/0s, 0s, and Giant 6s. The hands are blue steel with a center reservoir for luminous radium compound.

Value: $500–$700
Rarity: 7

Bottom left: The hinged case back. Bottom right: The original factory oversized crown.

ELGIN, SHADOW BOX DIAL IN BLACK ENAMEL

The watch shown below and on the next page sets the bar. It is the *only* size 0s black *enamel* dial watch you will find in this book. The other black Shadow Box Dials in this chapter are metal dials that have been refinished. Only two of these dials are known to exist. The font style is more European than American. The American versions of the shadow box tended to be block style, while this one is rounded. Waltham had a very similar black enamel dial, and those are rather common. The figures on this dial are golden in color and much of the original radium lum is still intact on the skeleton hands and numerals. The dial is signed ELGIN with a "USA" stamp just below. The movement is a 1916 grade 298 with 7 jewels, gilded finish, serial number 19331000. Elgin stopped manufacturing size 0s movements in late 1916, so this is one of the last ones that was produced. The movement is housed in a 10K yellow gold-filled "Giant" barrel case that was manufactured by the Illinois Watch Case Company. The case measures 42mm lug to lug, 33mm with the crown with 20mm lugs. It was thought for many years that the only black enamel dials that Elgin had available during the Great War were the Black Star Dials in the smaller size 3/0s; this one proves that some size 0s black enamel dials were produced, though obviously in small numbers.

Value: $2,500–$3,000
Rarity: 10+

This watch features a size 0s black enamel military dial and nice, wide 20mm lugs.

1916 ELGIN, SHADOW BOX DIAL

Relumed center reservoir military hands, Star Watch Case Company 14K gold-filled case with a hinged case back, factory crown, size 0s, 15 jewels, 1-piece 12mm leather strap.

Value: $500–$600
Rarity: 7

1916 ELGIN, SHADOW BOX DIAL

Relumed center reservoir military hands, Illinois Watch Case Company 10K gold-filled Giant cushion trench case, 10K gold-filled factory crown, size 3/0s, 15 jewels, 1-piece 12mm leather strap.

Value: $600–$700
Rarity: 8

1917 ELGIN, SHADOW BOX DIAL

Enameled center reservoir military hands, Illinois Watch Case Company sterling silver cushion trench case, factory crown, size 3/0s, 7 jewels, 12mm pigskin band.

Value: $575–$675
Rarity: 7

1917 ELGIN, SHADOW BOX DIAL

Enameled center reservoir military hands, "no-name" nickel case made by Wadsworth, lugs attached to the bezel, factory crown, size 3/0s, 7 jewels, 1-piece 16mm nylon strap.

Value: $400–$500
Rarity: 6

1918 ELGIN, SHADOW BOX DIAL

Relumed center reservoir military hands, 10K gold-filled Star Watch Case Company cushion trench case, factory crown, size 3/0s, 15 jewels, 16mm pigskin strap.

Value: $700–$800
Rarity: 9

1917 ELGIN, SHADOW BOX DIAL

Enameled center reservoir military hands, Illinois Watch Case Company sterling silver cushion trench case, factory crown, size 3/0s, 7 jewels, 1-piece 12mm nylon strap.

Value: $575–$675
Rarity: 7

1918 ELGIN, SHADOW BOX DIAL

Large numerals, enameled center reservoir military hands, Philadelphia 10K gold-filled case, replacement crown, size 3/0s, 7 jewels, 1-piece 14mm nylon strap.

Value: $450–$550
Rarity: 6

1911 ELGIN TRANSITION, SHADOW BOX DIAL

Refinished metal dial, relumed skeleton hands and dial numerals, American Watch Case Company sterling silver case, replacement crown, size 0s, 15 jewels, 1-piece 14mm nylon strap.

Value: $450–$550
Rarity: 6

1918 ELGIN, SHADOW BOX DIAL

Large numerals, enameled center reservoir military hands, "no-name" nickel case with an offset crown made by Wadsworth, factory crown, size 3/0s, 7 jewels, 3-piece 10mm bund strap.

Value: $500–$600
Rarity: 7

1918 ELGIN, SHADOW BOX DIAL

Large numerals, enameled center reservoir military hands, Philadelphia silverode case with an offset crown, factory crown, size 3/0s, 7 jewels, 3-piece 10mm bund strap.

Value: $500–$600
Rarity: 7

1918 ELGIN, SHADOW BOX DIAL

Large numerals, enameled center reservoir military hands, Philadelphia silverode case, factory crown, size 3/0s, 7 jewels, 3-piece 10mm bund strap.

Value: $450–$550
Rarity: 6

1917 ELGIN, SHADOW BOX DIAL

Refinished dial with a blue 12, enameled center reservoir military hands, "no-name" nickel case made by Wadsworth, lugs attached to the bezel, factory crown, size 3/0s, 7 jewels, 1-piece 16mm nylon strap.

Value: $400–$500
Rarity: 5

1917 ELGIN, SHADOW BOX DIAL

Metal dial, center reservoir military hands, Bates & Bacon 14K gold-filled semi-hermetic case with an offset crown, factory crown, size 3/0s, 7 jewels, 10mm leather band, 100 percent original unrestored condition with original radium lum on the dial.

Value: $600–$700
Rarity: 7

1918 ELGIN, SHADOW BOX DIAL

Large numerals, relumed skeleton hands, Philadelphia silverode case with an offset crown, factory crown, size 3/0s, 7 jewels, 16mm buffalo band.

Value: $500–$650
Rarity: 7

1917 ELGIN, SHADOW BOX DIAL

Relumed center reservoir military hands, 9K solid gold Keystone Watch Case Company with a hinged case back and a coin-edge bezel, factory crown, size 3/0s, 7 jewels, 3-piece 10mm bund strap.

Value: $900–$1000
Rarity: 9

1916 ELGIN, SHADOW BOX DIAL

Large numerals, enameled center reservoir military hands, "no-name" nickel case made by Wadsworth, factory crown, size 3/0s, 7 jewels, 16mm pigskin band.

Value: $500–$600
Rarity: 6

1915 ELGIN, SHADOW BOX DIAL

Enameled center reservoir military hands, Philadelphia silverode case, factory crown, size 0s, 7 jewels, 3-piece 10mm bund strap.

Value: $500–$600
Rarity: 6

1916 ELGIN, SHADOW BOX DIAL

Relumed center reservoir military hands, Giant 36mm Dennison sterling silver semi-hermetic case with a coin-edge bezel and case back, factory golden crown, size 0s, 15 jewels, 1-piece 12mm leather strap.

Value: $700–$800
Rarity: 9

1918 ELGIN, SHADOW BOX DIAL

Enameled center reservoir military hands, Philadelphia silverode case, replacement crown, size 3/0s, 7 jewels, 3-piece 10mm bund strap.

Value: \$375 - \$450
Rarity: 5

1917 ELGIN, SHADOW BOX DIAL

Refinished metal dial, relumed center reservoir military hands, rare set of 1917 Elgin military hands, "no-name" nickel case made by Wadsworth, factory crown, size 3/0s, 7 jewels, 1-piece 16mm nylon strap.

Value: \$500–\$700
Rarity: 7 (due to hands)

1918 ELGIN, SHADOW BOX DIAL

Large numerals, enameled center reservoir military hands, "no-name" nickel case made by Wadsworth, lugs attached to the bezel, factory crown, size 3/0s, 7 jewels, crown at 12 o'clock, 16mm buffalo band.

Value: \$450–\$550
Rarity: 6

1918 ELGIN, SHADOW BOX DIAL

Refinished dial, enameled center reservoir military hands, Philadelphia silverode case with an offset crown, factory crown, size 3/0s, 7 jewels, 3- piece 10mm bund strap.

Value: $500–$600
Rarity: 7

1917 ELGIN, SHADOW BOX DIAL

Enameled center reservoir military hands, Bates & Bacon 14K gold-filled semi-hermetic case with swivel lugs, crown at 12 o'clock, factory crown, size 3/0s, 7 jewels, 1-piece 14mm nylon strap.

Value: $900–$1,000
Rarity: 9

1918 ELGIN, SHADOW BOX DIAL

Relumed center reservoir military hands, wide lug configuration, Philadelphia silverode case, factory crown, size 3/0s, 7 jewels, 16mm buffalo strap.

Value: $500–$600
Rarity: 7

CHAPTER 9

THE CRYSTAL GUARDS

Among the most desirable Elgin Trench Watches are those that have crystal guards. The popular name or generalized term for this equipment on trench watches is "shrapnel guard," but the actual correct name is crystal guard, watch protector, or mesh guard. The same can be said about the term, "trench watch," which is simply a generalized term for wristwatches from the Great War era. I have not found a single reference to them being called "shrapnel guards" when they first came to market in 1914. Can you really imagine this small piece of tin actually stopping a piece of flying shrapnel while the wearer is trying to cross "no man's land"?

The name "crystal guard" says it all. They were designed to protect the watch crystal and face from cracking and from other light damage when in the field. A wide variety of design patents were filed for this optional equipment; I found no fewer than forty.

In 1917, at their peak of popularity, snap-on crystal guards could be bought for about 35 cents each, depending on the style. After the Great War ended, they could be had for only 20 cents each. Prices for these guards have gone up drastically in recent times! Just for the guard (not including the watch) be prepared to spend about $50 to $100 for an average design snap-on or strap-on model. For the more elaborate and rare designs, be prepared to pay from $200 up to a few thousand dollars.

In 1917, Joseph Fahys & Company came out with an "Armored Case" design that was a giant leap forward in technology. Instead of the crystal guard being strapped onto the case or snapped on over the case, the guard was actually part of the trench watch case itself. These are called "integrated crystal guard cases." The guard was snapped directly onto an outer ring around the bezel or, if the case was semi-hermetic, the guard was screwed into place.

Today, these integrated guard cases are highly sought after by collectors. Prices for integrated guard trench watches that have been fully restored are about $900 depending on the manufacturer, movement, and dial.

Over the years, some disreputable folks have manufactured fake copies of the snap-on and strap-on versions of the crystal guards. These fake versions sell for about $35. I estimate that nine out of ten "real" crystal guards had a "makers mark" or a "patent applied for mark" stamped into the guard. If you can't tell if a guard is authentic and you are buying online, ask an expert before making a purchase you will regret. Ask the seller to take a picture of the back of the guard and look for a maker's mark if there is any question about the authenticity.

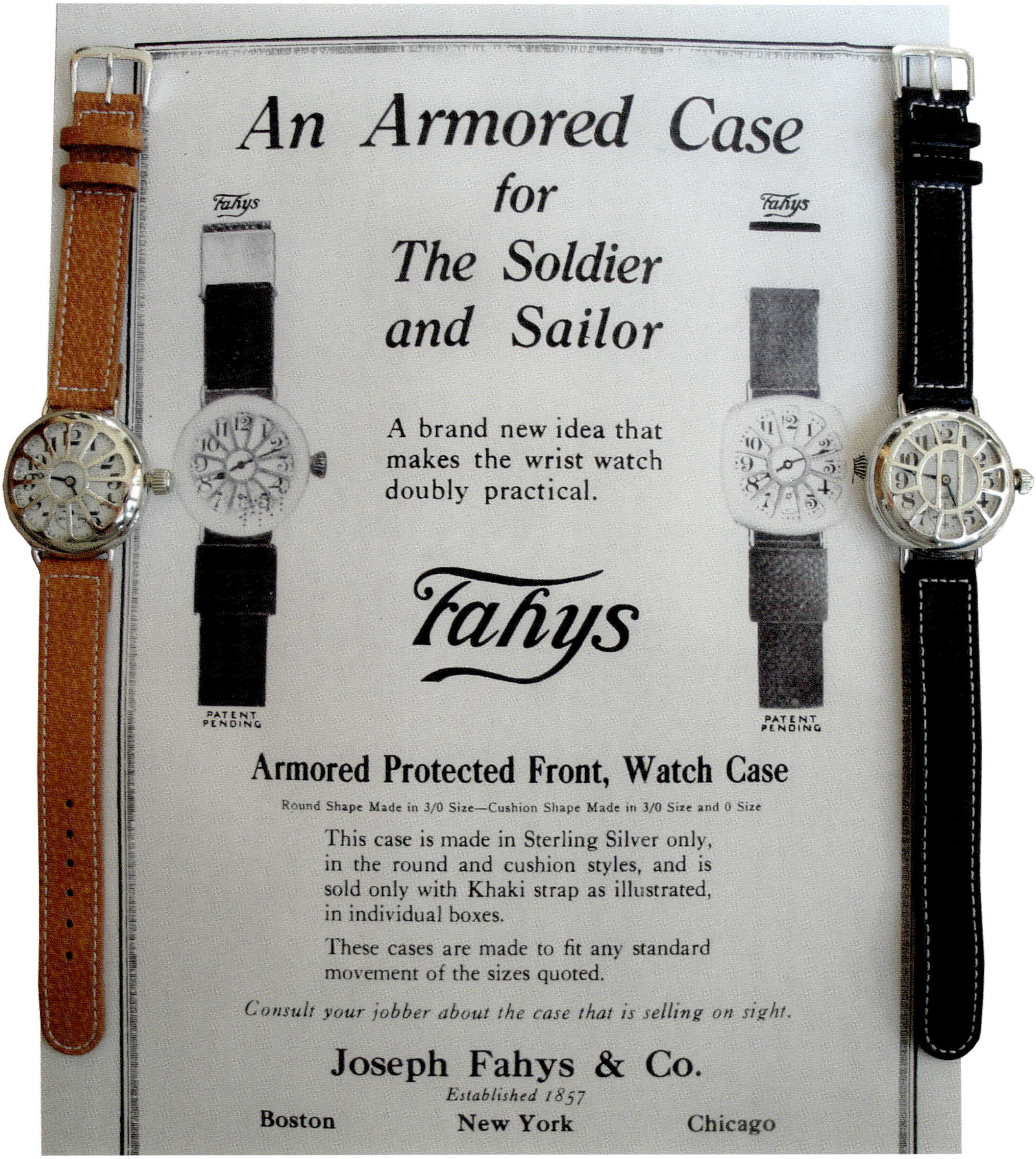

Elgin stopped manufacturing size 0s movements in very late 1916, so any Elgin movement you find in one of these size 0s Joseph Fahys & Company integrated crystal guard trench cases will pre-date the case, or a size-jumping movement spacer was used on a size 3/0s. The integrated guard cases from Fahys came out in September 1917; they were called "An Armored Case for the Soldier and Sailor."

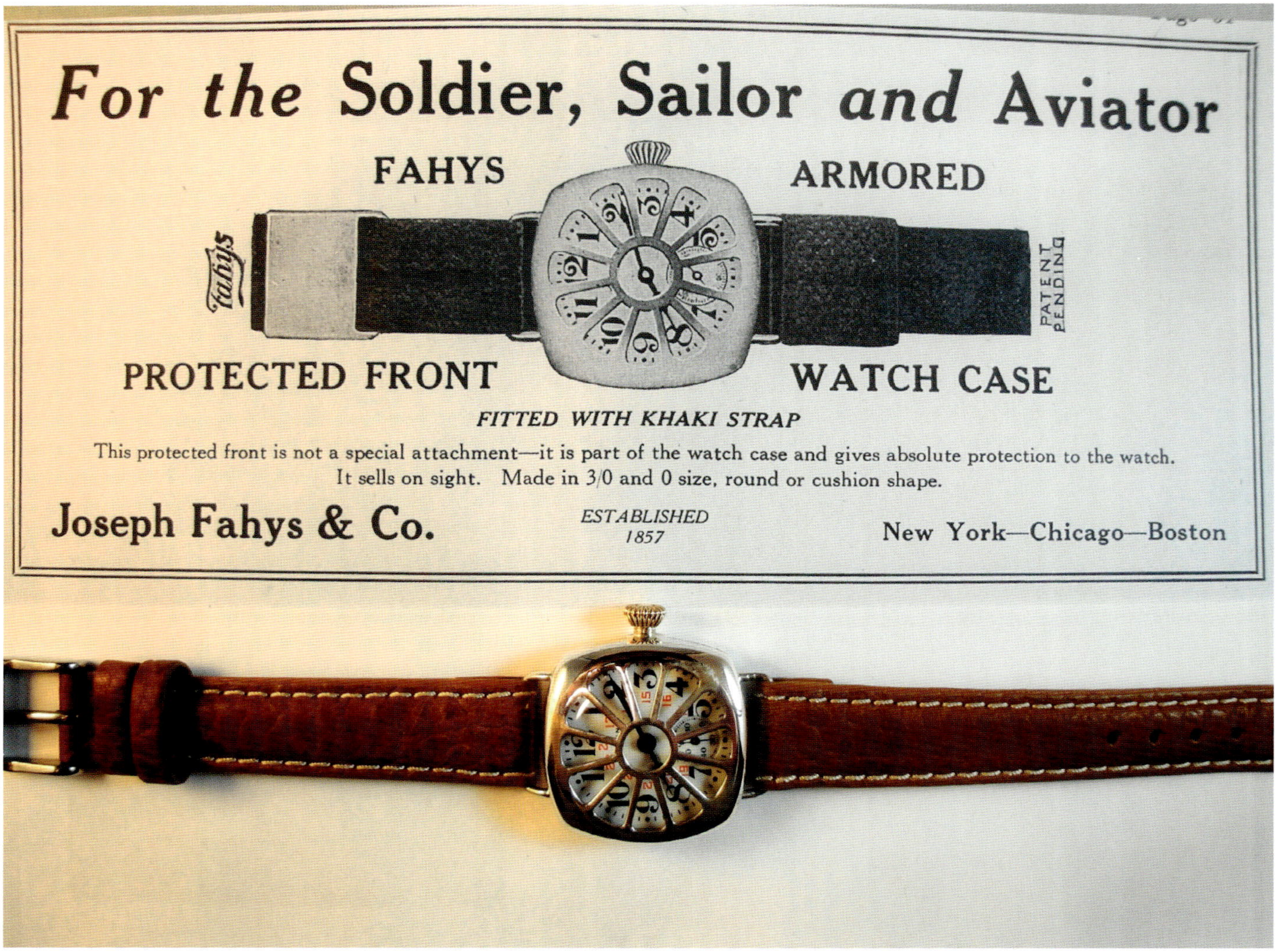

The 1917 size 3/0s Elgin shown here has a sterling silver Joseph Fahys & Company armored case with an integrated guard, cushion shape, BOLD Arabic dial, and thick, blue steel hands.

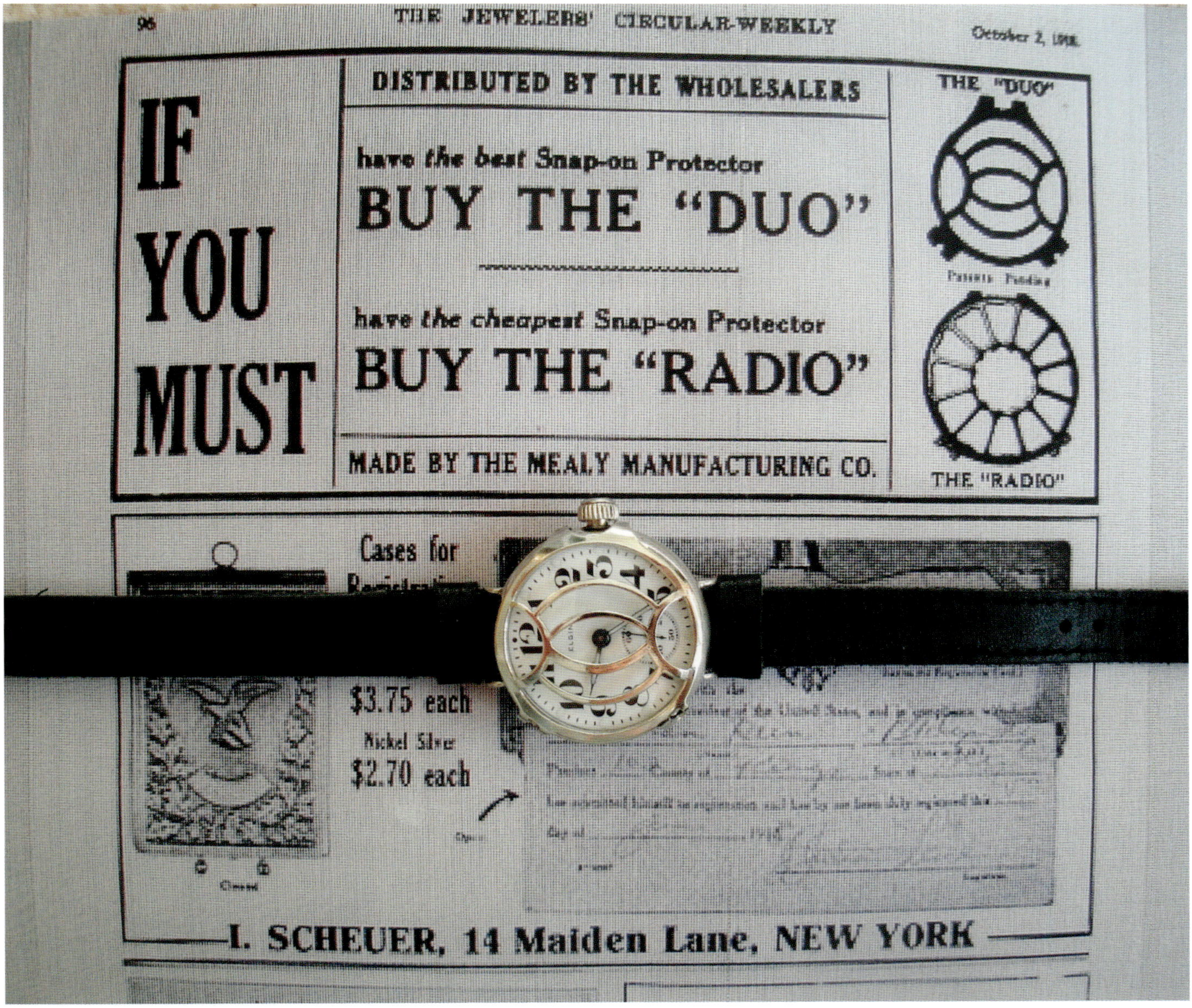

Some of the easiest guards to obtain these days were made by Mealy Manufacturing of Baltimore, Maryland. Their two most popular models were the "Duo" and the "Radio." This ad for Mealy Manufacturing is from October 2, 1918. The "Duo" design snaps on directly over the bezel and has a nice firm fit.

Here are two more "Duo" guards from Mealy Manufacturing.

These advertisements from the Mealy Manufacturing Company feature all of their guard designs.

Mealy Manufacturing advertisement, February 1918

Mealy Manufacturing advertisement, May 1918

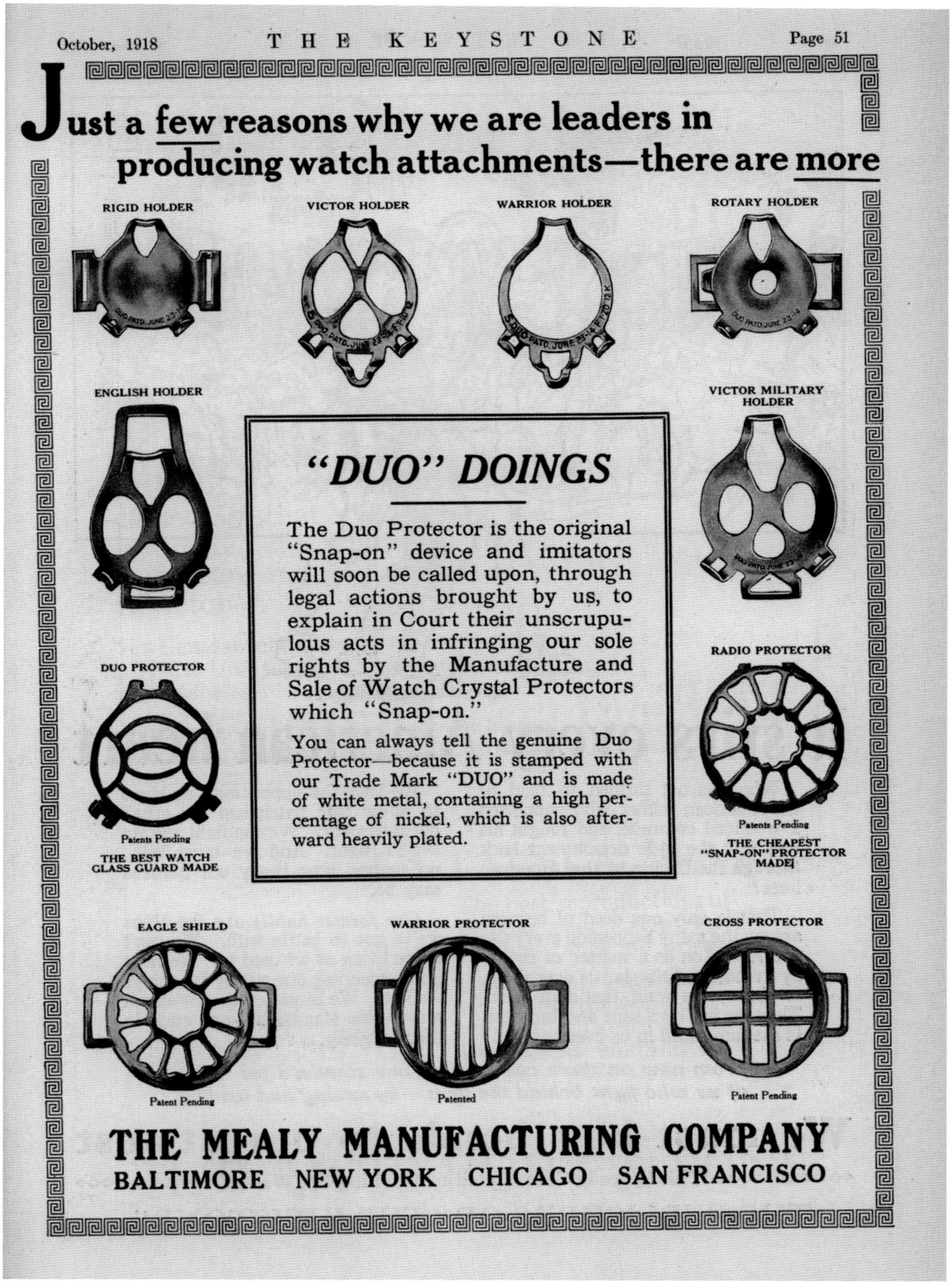

Mealy Manufacturing advertisement, October 1918

MEALY MANUFACTURING "DUO" CROSS PROTECTOR

The guard shown here was featured in the October 1918 ad on the previous page. What makes this protector rare is the metal content: it is made of sterling silver. It's the only sterling silver strap-on guard in this book. This guard has stamps for United States, English, and French patents. It is also stamped .925 for sterling silver content on one of the lug tips. This is a size 0s protector measuring 43mm lug to lug, 33mm going across, with a 10.5 lug diameter. This guard first came to market in late 1917 or early 1918.

Value: $400–$600
Rarity: 9

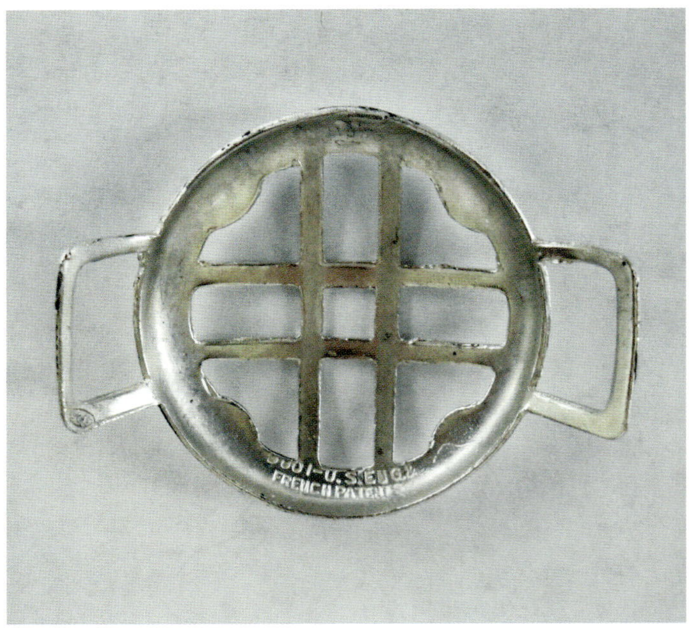

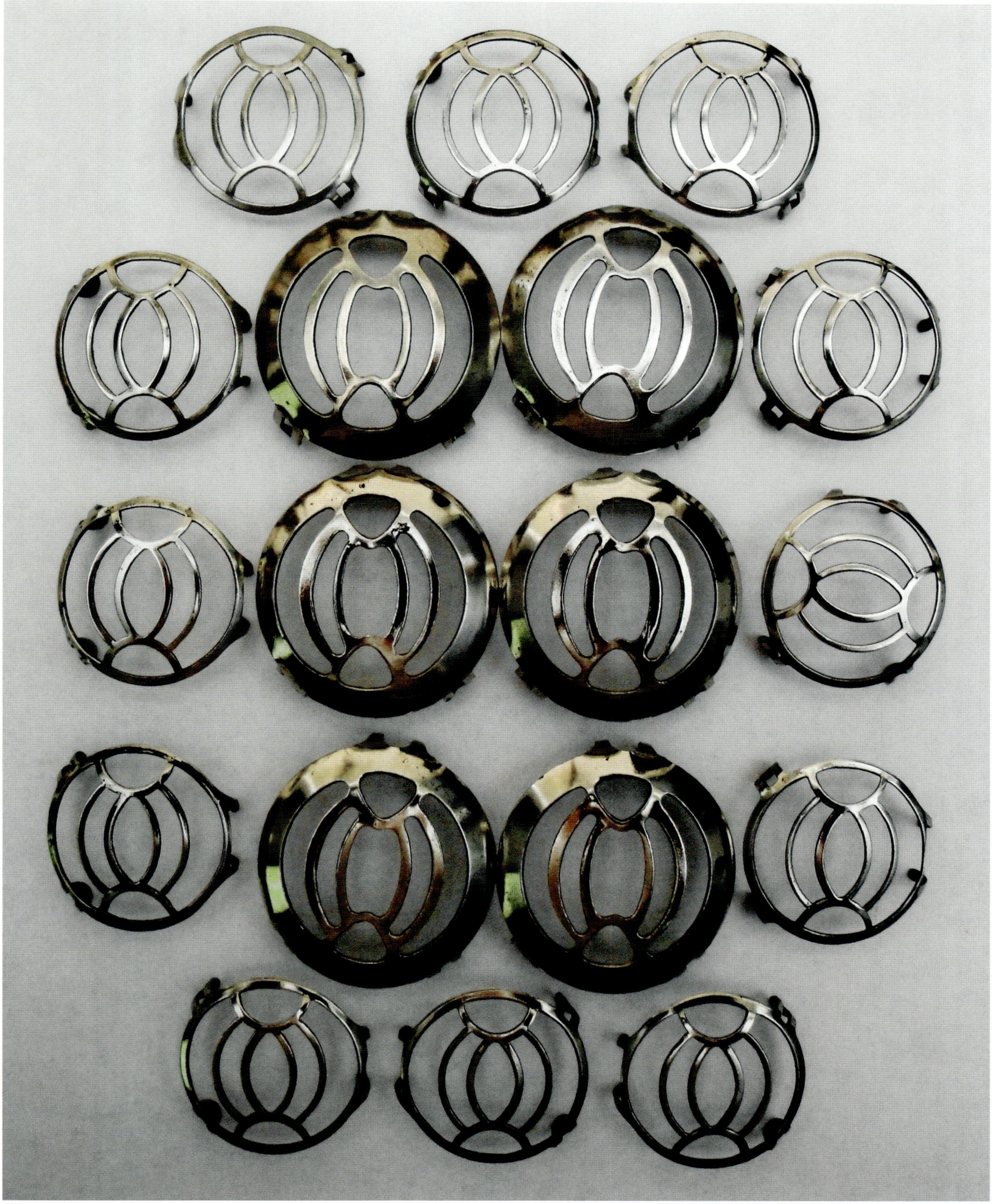

Size 0s and size 6s Mealy Manufacturing "Duo" protectors on parade.

Mealy Manufacturing was by far the largest manufacturer of crystal guards during the Great War, but they did have some good competition from S. Glagovsky of Haverhill, Massachusetts. The company's "Daisy" design was available in round or square shapes. This square shape would match up very nicely with the cushion trench cases made by the Illinois Watch Case Company and the Star Watch Case Company. Mealy Manufacturing did not make a square design, so Glagovsky had a slight step up on this one. The gold-plated versions are harder to obtain.

October, 1918 THE KEYSTONE Page 95

Sells Better

"DAISY"

WATCH PROTECTOR

Will Not Rust

The abundant sales and repeat orders on "Daisy" Protectors are ample proof of their superiority and practicability.

Made in Nickel, Bronze and Gun Metal.

Also made in 12 Kt. Rolled Gold Plate.

The "Daisy" Snap Protector is a wonderful crystal saver, and sells on sight.

We make a full line of the most durable and sightly leather watch straps for both men and women at a range of prices that permits the retailer to make a good profit. All Kitchener straps made in the finest quality pigskin and suede leather.

In ordering samples, designate width of center strap.

THE "DAISY" SNAP PROTECTOR

S. GLAGOVSKY

HAVERHILL, MASS.

Branch Office

HEYWORTH BLDG.

CHICAGO

We make these designs in a number of qualities

No. 111

LADIES' SAFETY WRIST WATCH STRAP

Made of the Finest Grades of Leather, in an assortment of colors, with Gold-Plated Trimmings, on individual tissue cards.

S. Glagovsky's "Daisy" protector.

GLAGOVSKY "DAISY" PROTECTOR, ROUND

Shown is a rare 12K rolled gold-plated, snap-on protector. It's the only 12K RGP version that I've been able to locate; the vast majority of them were made of nickel. This is the larger size 0s "Daisy," but they were also made in size 3/0s. This is a transition Elgin trench watch, as the movement pre-dates the case by about six to seven years. The 15 jewel size 0s movement is set inside a 10K gold-filled Philadelphia Victory case that still has the original factory gold-filled crown. The enamel RED 12 dial is in great condition without any hairline cracks. The RED 12 really stands out once the guard has been snapped into place. This snap protector is a bit of an oddity: it does not have a maker's mark or the patent date stamped underneath like the other Glagovsky protectors, but it is the real deal.

Value: \$1,000–\$1,500
Value for guard alone: \$300–\$400
Rarity: 10

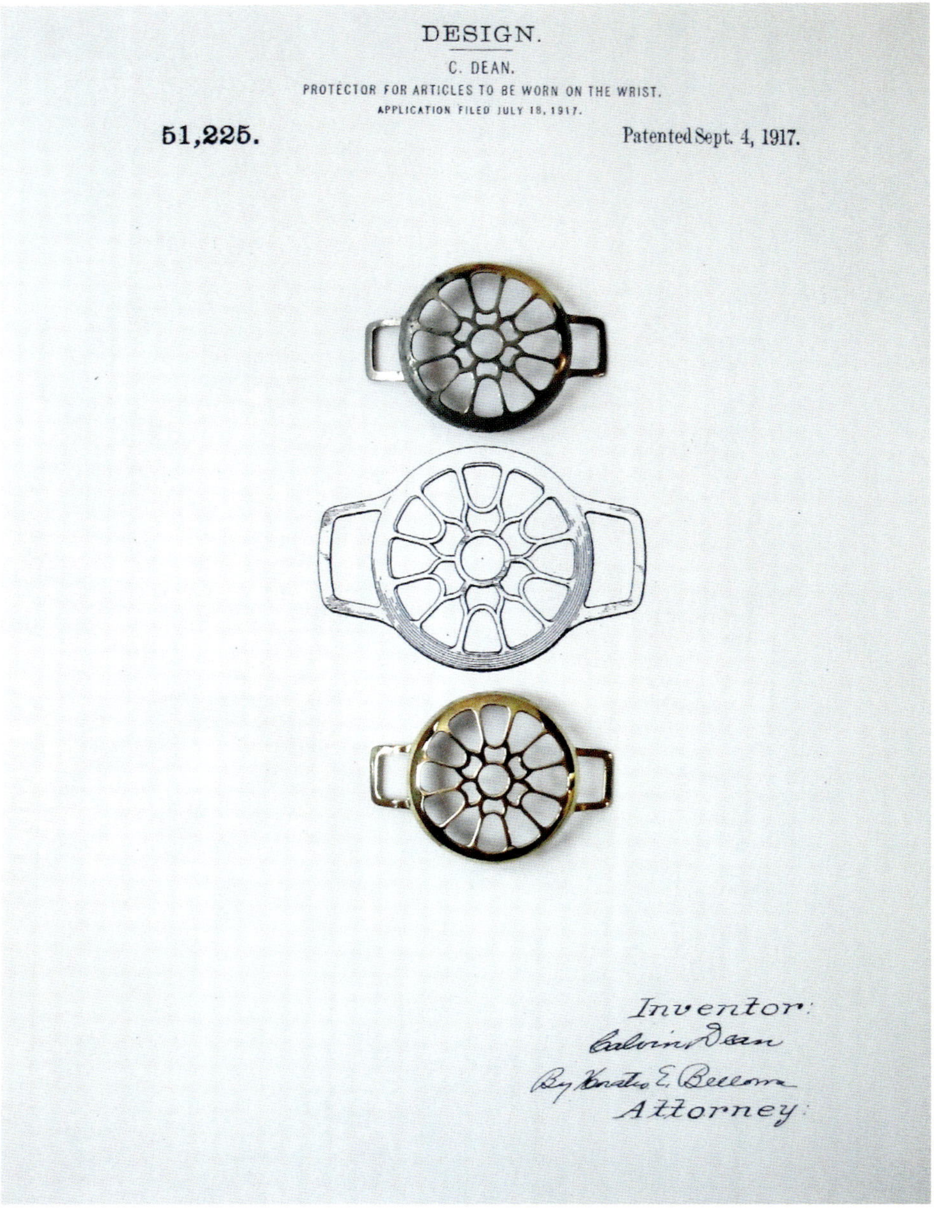

Here is the design patent for the "Daisy" watch protector. The patent application was filed on July 18, 1917, by Calvin Dean. The patent was granted on September 4, 1917.

Pictured at right are two identical "Daisy" watch protectors for a size 6s trench watch. The one on the left is stamped "PAT.APLD FOR," or Patent Applied For, meaning that it was manufactured between July 18, 1917, and September 3, 1917. This piece was one of the first of its kind to be manufactured. It was very common practice for a company to begin manufacturing a product as soon as the paperwork was filed with the United States Patent Office in Washington, D.C. The piece on the right is stamped "PAT'D 9-4-17" or Patented on September 4, 1917. It was manufactured after the patent was granted.

GLAGOVSKY "DAISY" PROTECTOR, CUSHION-SHAPE

This cushion-shaped "Daisy" watch protector is paired with a sterling silver cushion-shape trench watch case made by the Illinois Watch Case Company. The lugs on this guard are 16mm in diameter. The guard measures 44mm lug to lug and 32mm going across. The inside of the guard is stamped "PAT'D 9-4-17," the same as the round version. This 1919 Elgin features a Black Star Dial with a set of silver tone hands, grade 462 movement with 7 jewels, and is size 3/0s.

Value: $800–$1,000
Rarity: 9

Mealy Manufacturing also made a crystal guard for pocket watches. Called the "Warrior," it was designed with a holder that hung from a pants belt. I've seen a few variations of this design manufactured for size 6s open face pocket watches. Mealy also made protectors that could convert pocket watches to wristwatches: wrist straps could be looped through the protectors' lugs..

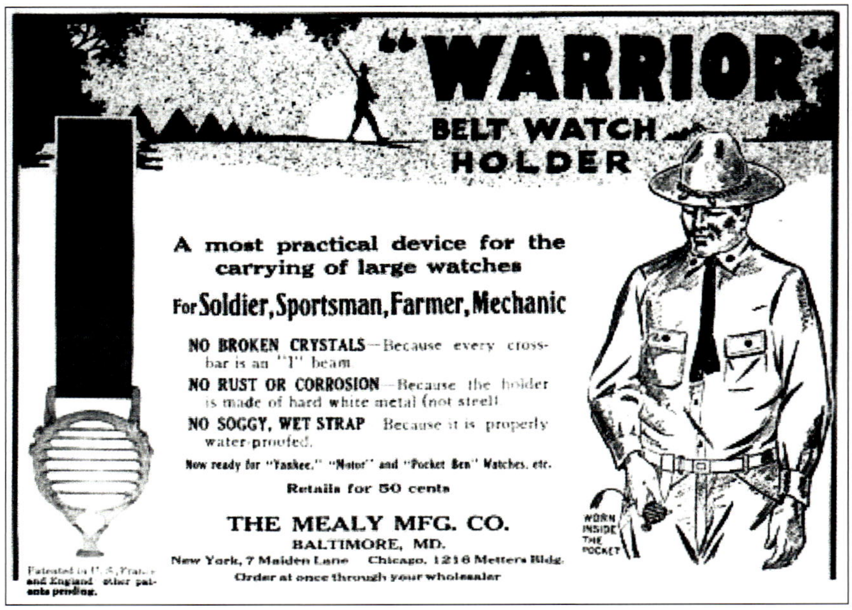

Some of the very first guards were universal: they worked with either Swiss or American wristwatch cases. They would completely surround the case, somewhat like a clam shell with a spring-loaded hinge. The design shown in the ad above and at left is called the P.T.O (Please Turn Over) protector. It was one of the very first guards ever manufactured. It was designed by Charles Schierwater of Liverpool, England, in October 1914. It had an "on-duty" and "off-duty" side. While "on duty," the watch face would be completely covered so soldiers wouldn't give away their position in the field. With the "off duty" side, the watch face was visible.

1917, ELGIN GIANT SIZE 6S WITH P.T.O. PROTECTOR

The Please Turn Over guards were available in four sizes. The example shown below is the largest and would fit a size 6s trench watch. This guard measures 45mm across and 38mm top to bottom. The insides of the guards are stamped with patent number 20696–14, the 14 is for the year 1914. These guards are quite rare; prices for the guard alone (without the watch) are $150 to $350. Pictured with this guard is a Giant size 6s 1917 Elgin trench watch. Shadow Box enamel military dial with the correct 1917 hands, 10K gold-filled Napoleon case with swivel lugs made by the Illinois Watch Case Company. It features a grade 286 movement with 7 jewels. The case measures 50mm lug to lug, 37mm without the original factory crown with a 14mm lug diameter.

Value: $1,700–$2,000
Rarity: 10+

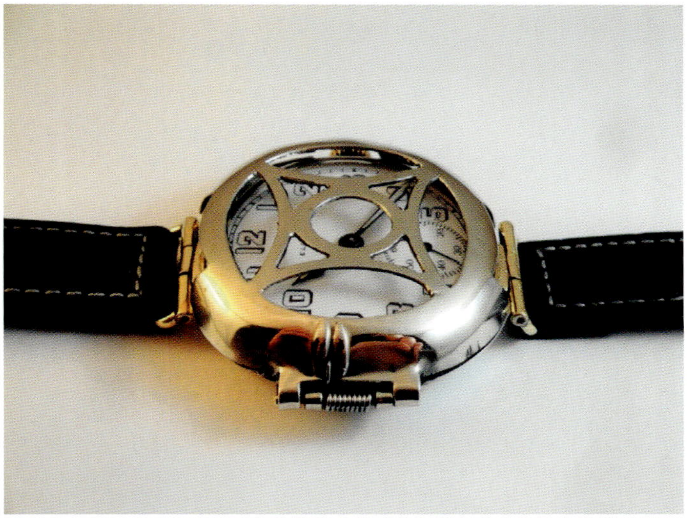

Top: Close-ups show the spring-loaded hinge of the P.T.O protector.
Bottom left: "On-duty" side of protector. **Bottom right:** "Off-duty."

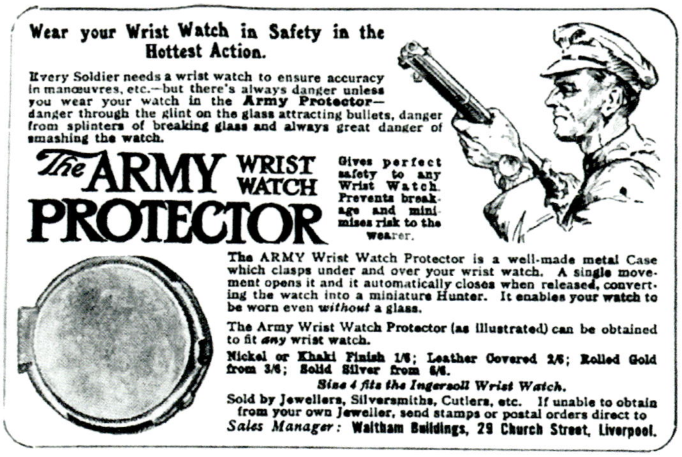

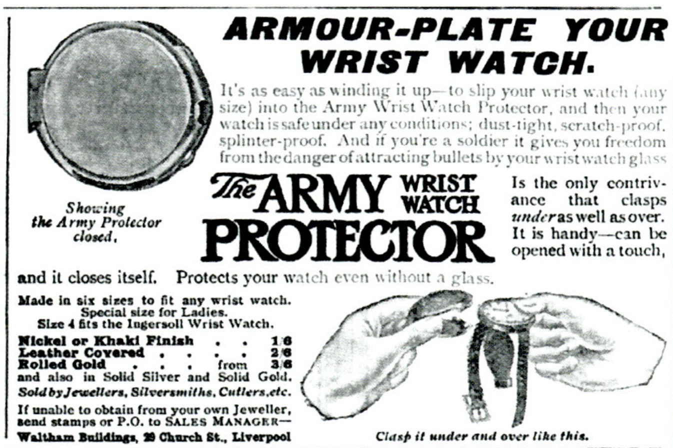

In addition to his P.T.O. protector, Charles Schierwater also had another guard design. This one did not have a see-through grill. It was solid on both sides. In order to read the time, one would have to lift the front cover up.

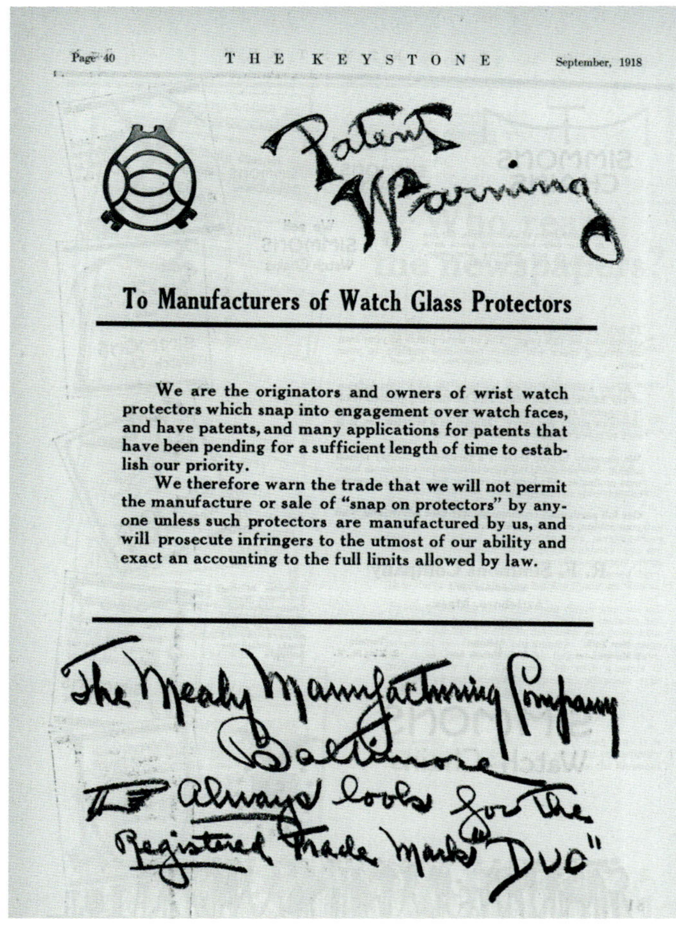

The competition among companies manufacturing snap-on watch profectors was fierce.

Below are four really great crystal guards. The one at far left is extremely rare.
From left to right.
Made by Mealy Manufacturing, the "Warrior," February 1918.
Made by Mealy Manufacturing, the "Duo," October 1917.
Made by Mealy Manufacturing, name unknown, date unknown.
Designed by Louis Jacot of Birmingham England, January 1918.

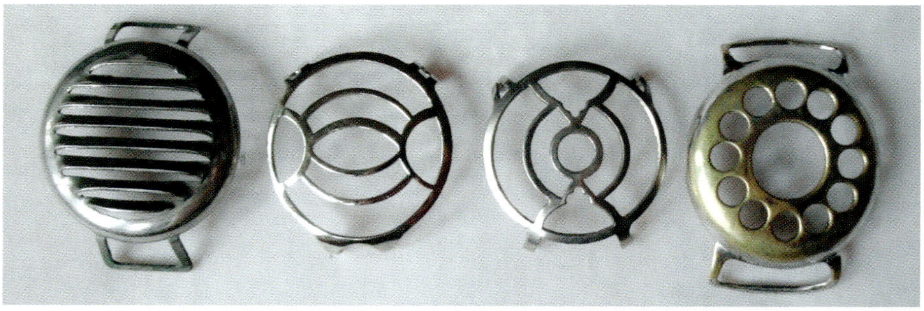

An "Improved Mesh-Guard" with a khaki finish for a size 6s trench watch, original packaging.

Pictured above are two fake crystal guards that are recent reproductions, probably made in China. These are *not* genuine, and they are not antiques; they are poor copies of a February 3, 1916 "pinwheel" design invented by Frank Farr of Quebec, Canada, patent #1199256. There are many dealers who try to sell these as genuine guards.

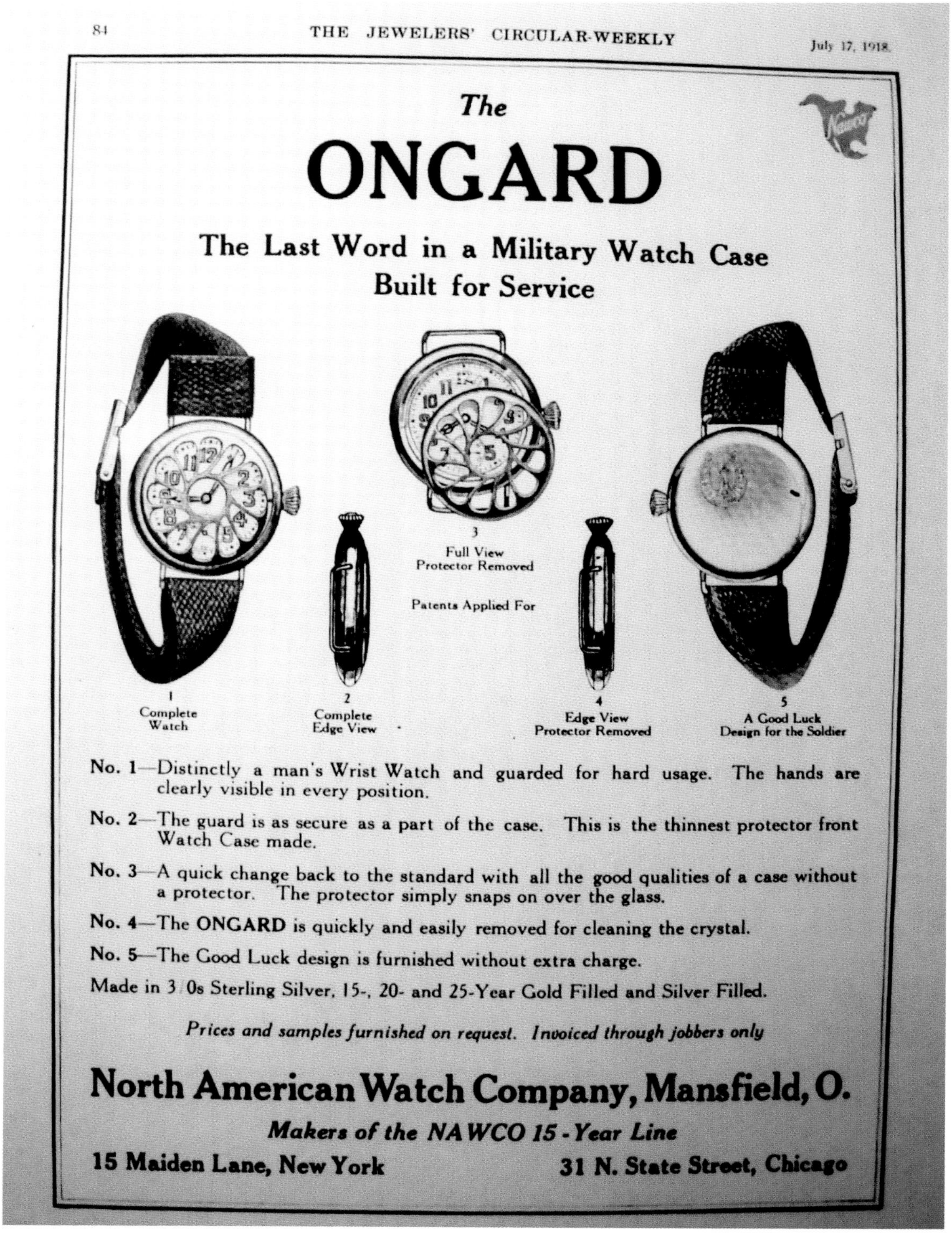

84 THE JEWELERS' CIRCULAR-WEEKLY July 17, 1918.

The

ONGARD

The Last Word in a Military Watch Case Built for Service

3
Full View
Protector Removed

Patents Applied For

1
Complete Watch

2
Complete Edge View

4
Edge View
Protector Removed

5
A Good Luck Design for the Soldier

No. 1—Distinctly a man's Wrist Watch and guarded for hard usage. The hands are clearly visible in every position.

No. 2—The guard is as secure as a part of the case. This is the thinnest protector front Watch Case made.

No. 3—A quick change back to the standard with all the good qualities of a case without a protector. The protector simply snaps on over the glass.

No. 4—The **ONGARD** is quickly and easily removed for cleaning the crystal.

No. 5—The Good Luck design is furnished without extra charge.

Made in 3/0s Sterling Silver, 15-, 20- and 25-Year Gold Filled and Silver Filled.

Prices and samples furnished on request. Invoiced through jobbers only

North American Watch Company, Mansfield, O.

Makers of the NAWCO 15-Year Line

15 Maiden Lane, New York **31 N. State Street, Chicago**

The North American Watch Case Company introduced integrated crystal guard watch cases in early 1918. The were available in sterling silver, silver-filled, and gold-filled. NAWCO was a much smaller company than Fahys, and fewer of these cases exist. One nice feature of these cases is the "good luck" stamp on the outside of the case back: it was a horseshoe with "USA" in the middle.

NAWCO "ONGARD" MILITARY WATCH CASE WITH INTEGRATED CRYSTAL GUARD

This case (shown below and on the next page) with its offset crown is one of the rarest American trench watches from the Great War. Only a few are known to exist. The 4-piece snap-together, sterling silver case still has the original sterling silver factory crown. This case is very large for a size 3/0s. It measures 38mm lug to lug, 33.5mm without the crown and has 16mm lugs. It features a 1918 grade 463 open face Elgin movement with 7 jewels. Nearly all of the original factory radium lum is still intact on the dial and hands. This watch still has the original glass crystal; it now has a purple hue from radium burn. This is one of the very first "ONGARD" cases manufactured by the NAWCO. The inside of the case back reads "PAT. APPL'D FOR."

Value: $3,000–$4,000
Rarity: 10+

1918 Elgin with sterling silver "ONGARD" case with an offset crown and an integrated crystal guard.

The sterling silver case has four pieces, including the integrated guard.

"GIANT" 0S CASE WITH INTEGRATED CRYSTAL GUARD

This Elgin trench watch (shown here and on the next page) is simply as good as it gets, for several reasons. This is a 10K yellow gold-filled "Giant" case that was manufactured by the Illinois Watch Case Company in 1918. It has swivel lugs with an integrated crystal guard. The bezel and case back are threaded, so they screw on and off instead of popping on and off, a far superior design. This combination of 10K gold fill, swivel lugs, and threaded case with an integrated crystal guard is incredibly rare. The case is in superb condition without any gold fill wear spots. The case is quite large for a size 0s, measuring 45mm lug to lug, 34.5mm without the gold-filled factory crown with a 14mm lug diameter. This is a 3-piece case with the crystal seated underneath the bezel. A 14mm 2-piece genuine pigskin strap has been installed.

The dial on this Elgin (right) is as rare as the case. This is a BOLD Arabic large font RED 12 enamel dial with a "USA" stamp just below the Elgin signature. Usually when you see a RED 12 in this style, the font size is much smaller. Compare it with the Arabic font RED 12 enamel dial (left); the difference is quite clear. .

The threaded case back is stamped "Illinois Watch Case Company, -ELGIN-" and "Giant" indicating that it is 10K gold-filled. The case serial number is 5642319.

Elgin stopped manufacturing size 0s movements in late 1916. This case was manufactured in 1918, so any movement would pre-date the case, unless a size-jumping spacer was used on a size 3/0s movement. The movement installed in this watch is one of the best Elgin ever produced. This is a size 0s movement with 17 jewels, grade 263. The early grade 263 movements, like this one, had a solid gold train. In later years, they only had a solid gold center wheel. The serial number on this movement is 9141270, dating it back to 1901. The solid gold raised jewel chatons and demaskeening pattern on this movement are a thing of beauty.

Value: $4,000–$6,000
Rarity: 10++

Details, clockwise from top right: Grade 263 movement with solid gold train; threaded, integrated crystal guard; swivel lugs.

1917 ELGIN WITH "LATH" SHRAPNEL GUARD

The trench watch shown here features one of the most sought-after and rare guards from the Great War. It was designed by Harry Daw of High Wycombe, England, in January 1916. This is a high-quality guard of solid nickel construction; most guards were just plated. The silverode watch case was made by the Philadelphia Watch Case Company, and it still has the original factory crown. This is a size 0s case but the movement is a size 3/0s; a size-jumping spacer was used during assembly. Grade 417 movement with 7 jewels dating back to 1917, serial number 20221641. This watch still has the original 1917 model year lum reservoir hands. In 1917, the hour hand had a round lum reservoir; in later years the lum reservoir was rectangular. The enamel dial has shadow box numerals and an outer chapter of minute markers. It's somewhat rare to find a "USA" stamp just below the Elgin signature in these dials. The shrapnel guard measures 46mm lug to lug, 35.5mm going across with a 10mm lug diameter. The case measures 39mm lug to lug, 32mm without the crown with a 10mm lug diameter.

Value: $1,300–$1,500
Rarity: 10

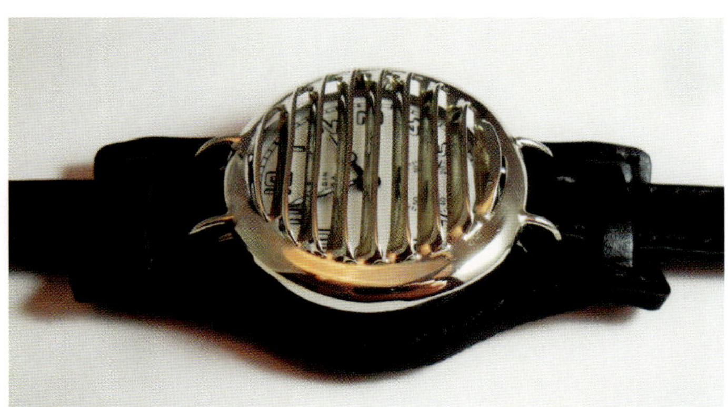

There are two main differences between the "Lath" design of Harry Daw and Mealy Manufacturing's "Warrior" design. The Daw design lugs were soldered into place on top the bezel while the Mealy design lugs were stamped or cut directly as part of the guard. The Daw design has nine lathes going across the top, while the Mealy design has eight. The maker's marks on the inside are different as well. The Daw design is stamped: 3 PROV PATE. The Mealy design is stamped: DUO 2 U.S. ENGLISH & FRENCH PATENTS.

Harry Daw's "Lath" design.

A top view of the "Lath".

Mealy Manufacturing's "Warrior" design.

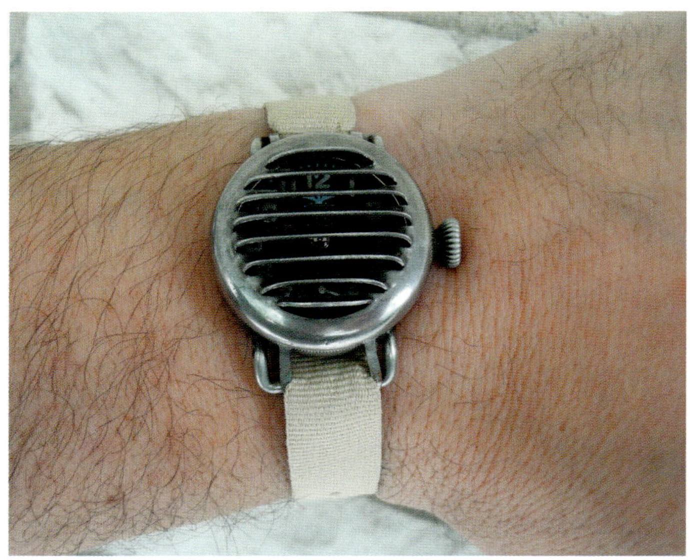

The top view of the "Warrior."

Patent materials for the Daw design.

UNITED STATES PATENT OFFICE.

HARRY DAW, OF HIGH WYCOMBE, ENGLAND.

PROTECTOR FOR WRIST-WATCHES.

1,249,906. Specification of Letters Patent. **Patented Dec. 11, 1917.**

Application filed January 17, 1916. Serial No. 72,542.

To all whom it may concern:

Be it known that I, Harry Daw, a subject of the King of Great Britain, residing at 7 Oxford street, High Wycombe, in the county of Buckinghamshire, in England, have invented a new and useful Improved Protector for Wrist-Watches, of which the following is a specification.

The invention relates to an improved protector for wrist watches.

In the accompanying drawings Figure 1 is a top or edge view of the disk, Fig. 2 is a front elevation or face view of the disk, and Fig. 3 is a back elevation of the disk before the laths are bent to the required angle with the disk body.

My invention consists of a metal disk 1, suitably domed to cover a watch, with a series of bars or laths *a* cut across the center with perforations *b*, each bar *a* to be set at an angle after the manner of the laths of a Venetian blind, so that by tilting the watch a view of the dial and hands is obtained. The metal disk *a* is to be held in position by a loop *c* or slot on each side thereof, through which the strap of the watch would pass.

I claim:

A metal disk suitably domed to cover a watch, having a series of slits cut across the center forming bars or laths, said bars being bent at an angle with the disk body, so that by tilting the watch a view of the dial and hands is obtained, the said disk being provided with loops through which the strap of the watch passes.

HARRY DAW.

Witnesses:
E. W. Bayly,
W. E. P. Bayly.

Copies of this patent may be obtained for five cents each, by addressing the "Commissioner of Patents, Washington, D. C."

H. DAW.
PROTECTOR FOR WRIST WATCHES.
APPLICATION FILED JAN. 17, 1916.

1,249,906. Patented Dec. 11, 1917.

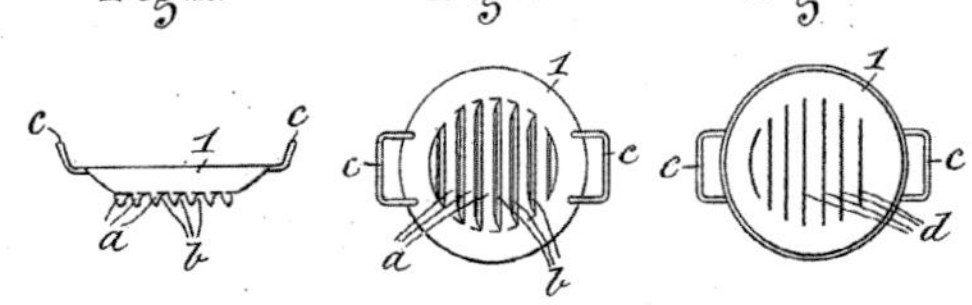

Witnesses.
E. W. Bayly
W. E. P. Bayly

Inventor:
Harry Daw
By John Pitt Bayly
attorney

"LATH" PROTECTOR FOR "GIANT" SIZE 6s

Harry Daw's "Lath" design was also available for the "Giant" size 6s trench watches. I recently obtained one of these very rare guards, which is shown below at right (at left is the Mealy size 0s "Duo" crystal guard). It measures 45.97mm lug to lug and 38.64mm across. This larger size guard has nine lathes, while Daw's 0s size "Lath" has nine. The larger versions of guards and watches are historically more valuable than their smaller counterparts, This is the only size 6s "Lath" I have been able to locate; it fits a size 6s Philadelphia trench watch case with the narrow 10mm lugs. Very few of the Philadelphia cases exist with the smaller 10mm lugs, but they can be found. If you put it on the more common Philadelphia size 6s case with 16mm lugs, it simply would not look right.

Price for the guard: $3,000
Rarity: 10+

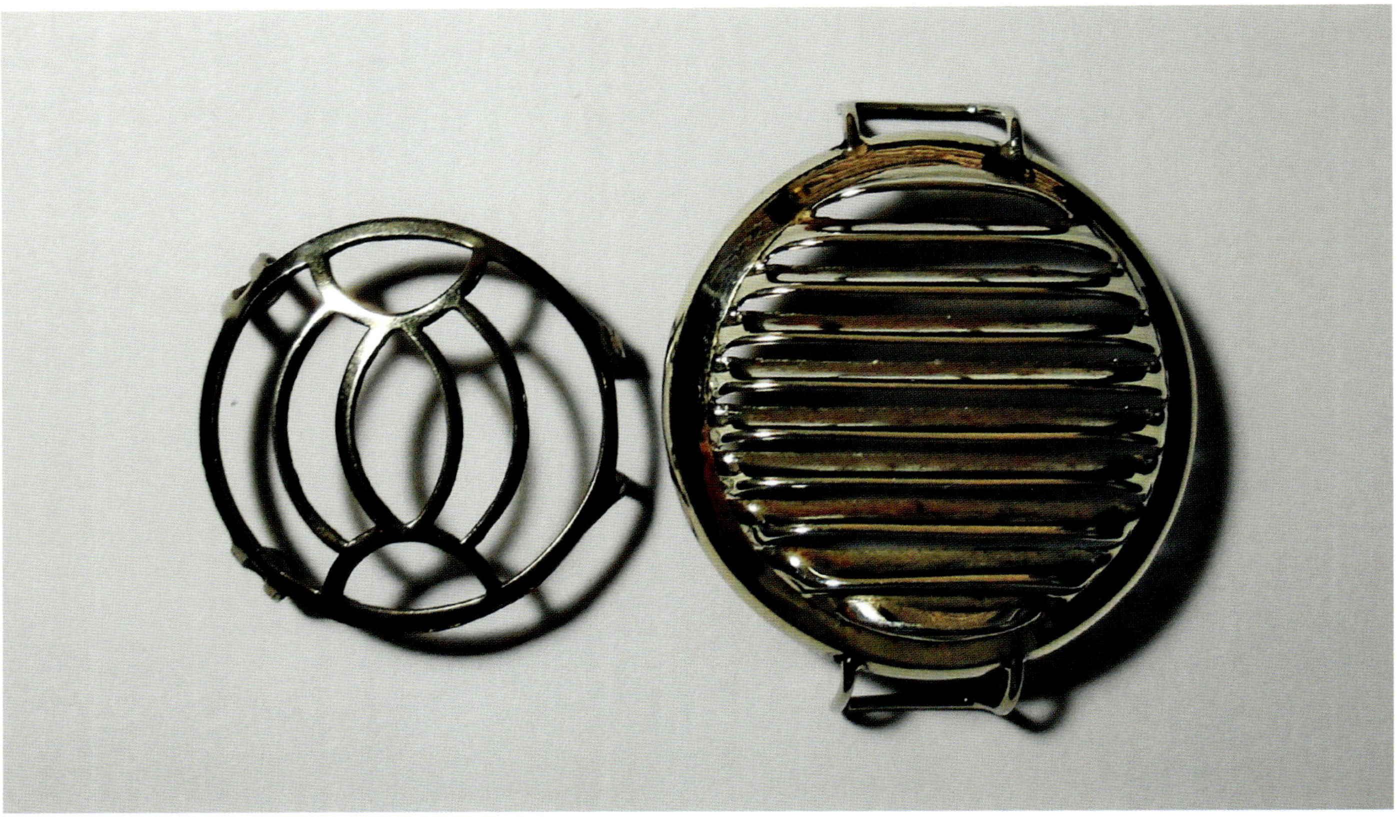

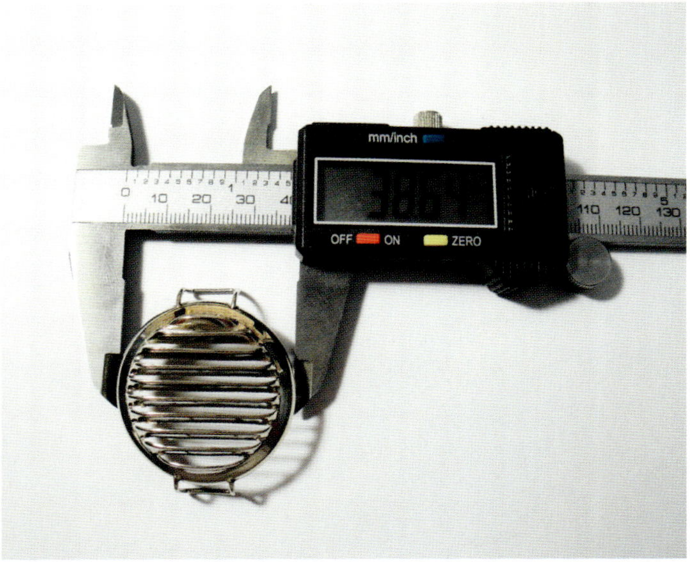

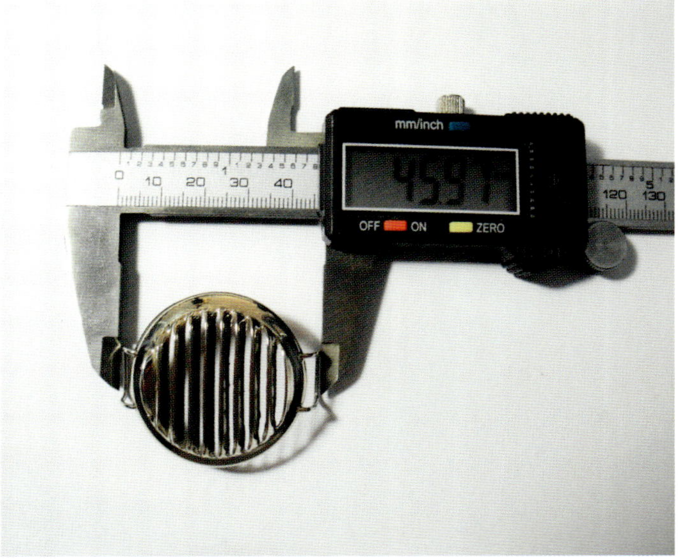

1918 ELGIN, BLACK STAR DIAL, BLACK FINISH FAHYS ARMORED CASE

Starting on October 24, 1916. the United States Army required that all watch cases be black in color (General Specification, No. 579-D). Nickel and silverode cases could be oxidized for the required black case finish. But sterling silver cases had to be painted. Very few pristine examples of the black finish can be found today. I believe that the U.S. Army changed the general specification sometime around mid-1918, allowing sterling silver cases to be used, but according to sources at the National Archives, this document has unfortunately been lost in time. However, the pieces of the puzzle can be put back together from advertisements and documented watch cases that were used by the military from specific years, like 1917 and 1918.

The restoration of the watch shown below and on the next page will undoubtedly spark controversy between the watch purist who thinks no watch should ever be restored and the watch collector who likes antique watches fully restored and looking brand new. I restored this Elgin to show what the watch looked like when it was brand new. Requiring a dull black finish on a watch case was a pretty good idea. A soldier would not want to give an enemy sniper from across no-man's land an easy target, like a shiny watch case. This watch is size 3/0s, 7 jewels, grade 463.

Value: $1,200–$1,500
Rarity: 9

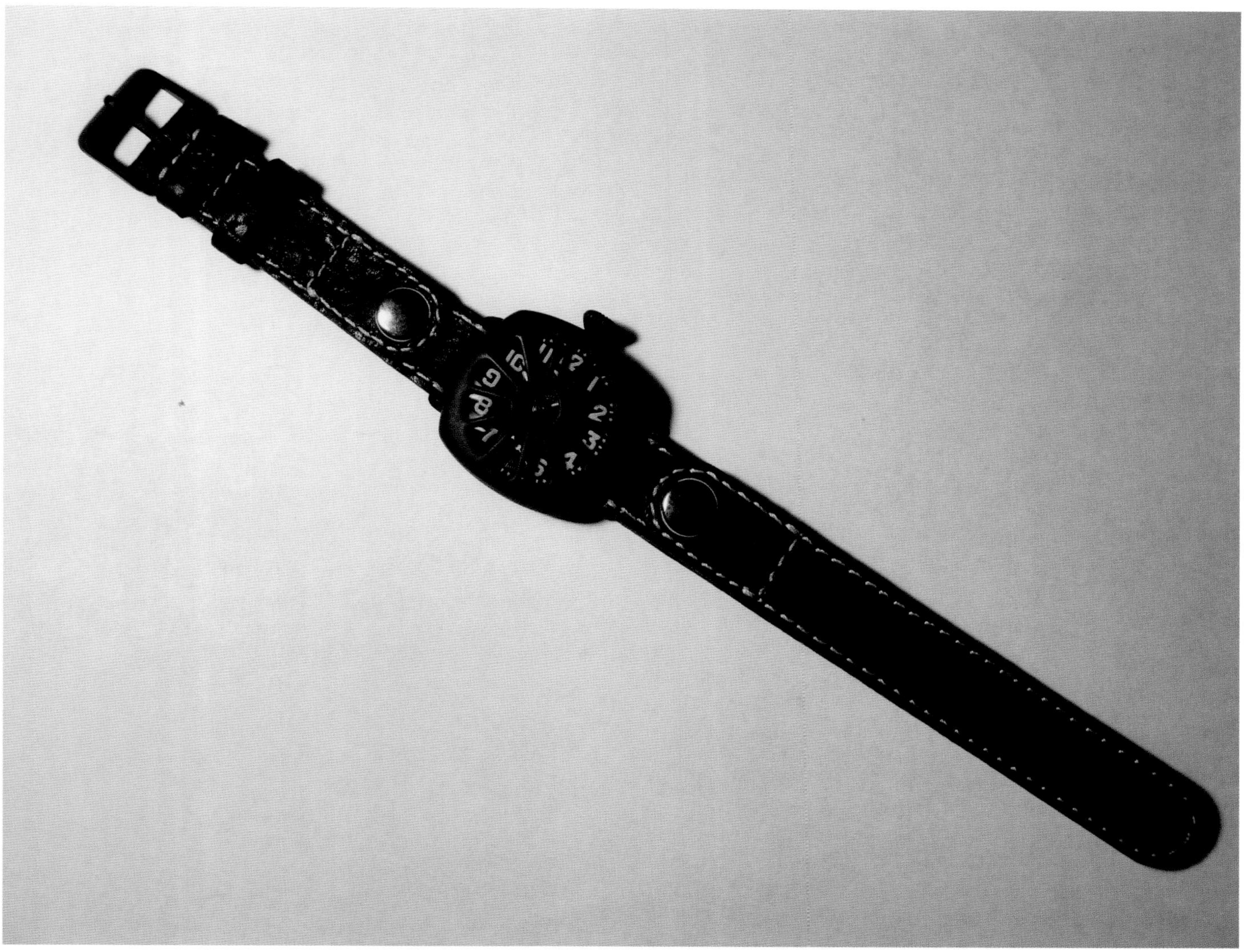

This restored Fahys "Armored" case with an integrated guard has been painted black. The entire case's dull black finish—from crown to buckle—are per the U.S. Army's wartime specifications.

1918 ELGIN, BLACK STAR DIAL

Enameled center reservoir military hands, Fahys sterling silver "Armored" case with integrated crystal guard, replacement 1915, 17 jewel, size 3/0s movement, (1 of only 500 made), replacement crown, 16mm nylon strap.

Value: $1,500–$1,750
Rarity: 10+ (due to the rare movement)

1918 ELGIN, SHADOW BOX DIAL

Relumed center reservoir military hands, Fahys sterling silver "Armored" case with integrated crystal guard, size 3/0s, 7 jewels, replacement crown, 16mm buffalo band.

Value: $1,000 - $1,200
Rarity: 9

1918 ELGIN, BLACK STAR DIAL

Relumed center reservoir military hands, Philadelphia silverode case, Mealy Manufacturing "Duo" snap-on crystal guard, factory crown, size 3/0s, 7 jewels, 1-piece leather strap.

Value: $600–$750
Rarity: 8

1916 ELGIN, LARGE NUMERAL SHADOW BOX DIAL

Enameled center reservoir military hands, "no-name" solid nickel case made by Wadsworth, Mealy Manufacturing "Duo" snap-on crystal guard, factory crown, size 3/0s, 7 jewels, 16mm pigskin band.

Value: \$550–\$650
Rarity: 7

1918 ELGIN, GIANT 6s, SHADOW BOX DIAL

Relumed center reservoir military hands, Illinois Watch Case Company nickel case with swivel lugs and a hinged case back, 37mm x 52mm, size 6s, khaki finish mesh guard, 7 jewels, 1-piece leather strap.

Value: \$1,500–\$1,700
Rarity: 10+

1918 ELGIN, BLACK STAR DIAL

Relumed skeleton military hands, Fahys sterling silver "Armored" case with integrated crystal guard, replacement crown, size 3/0s, 7 jewels, 16mm pigskin band.

Value: \$1,000–\$1,200
Rarity: 9

1917 ELGIN, WHITE STAR DIAL

Metal dial, center reservoir military hands, Philadelphia sterling silver semi-hermetic case with an offset crown, "Depose" crystal guard, factory crown, size 3/0s, 7 jewels, 3-piece 10mm bund strap.

Value: $700–$800
Rarity: 8

1918 ELGIN, BLACK STAR DIAL

Skeleton military hands, Fahys oresilver semi-hermetic case with a coin-edge bezel, "Lath" crystal guard, factory crown, size 3/0s, 7 jewels, 1-piece nylon strap.

Value: $1,000–$1,200 (due to very rare crystal guard)
Rarity: 10

1912 ELGIN TRANSITION, BOLD ARABIC DIAL

Moon hands, Fahys "Armored" case with an integrated crystal guard, sterling silver, replacement crown, size 0s, 7 jewels, 16mm pigskin band.

Value: $900–$1,000
Rarity: 9

1914 ELGIN TRANSITION, BOLD ARABIC DIAL

Moon hands, Fahys "Armored" case with an integrated crystal guard, solid nickel, replacement crown, size 0s, 7 jewels, 16mm pigskin band.

Value: $900–$1,000
Rarity: 9

1914 ELGIN, BOLD ARABIC DIAL

Relumed center reservoir military hands and factory-drilled hour marker lum dots, "no-name" nickel case made by Wadsworth with a hinged case back, Mealy Manufacturing "Duo" crystal guard, factory crown, size 0s, 7 jewels, 1-piece leather strap.

Value: $600–$750
Rarity: 8

1915 ELGIN, SHADOW BOX DIAL

Relumed center reservoir military hands, solid nickel 34mm semi-hermetic Illinois case with swivel lugs, Mealy Manufacturing "Duo" crystal guard, factory crown, size 0s, 7 jewels, 14mm green canvas strap.

Value: $800–$1,000
Rarity: 9

1917 ELGIN, BOLD ARABIC DIAL

Relumed center reservoir military hands and factory drilled hour marker lum dots, Philadelphia silverode case, 42mm x 38mm, Mealy Manufacturing "Duo" crystal guard, factory crown, size 6s, 7 jewels, 2-piece genuine buffalo strap.

Value: $1,700–$2,000
Rarity: 10+

1917 ELGIN, BOLD ARABIC DIAL

Relumed skeleton hands and factory-drilled hour marker lum dots, Illinois solid nickel case, 48mm x 38mm, khaki mesh guard, replacement crown, size 6s, 15 jewels, 1-piece leather strap.

Value: $1,300 - $1,500
Rarity: 10

1917 ELGIN, 24-HOUR MILITARY DIAL

Thick blue steel hands, Fahys sterling silver "Armored" case with an integrated crystal guard, factory crown, size 3/0s, 7 jewels, 16mm genuine buffalo strap.

Value: $1,300–$1,600
Rarity: 10

1913 ELGIN, BOLD ROMAN DIAL

Thin spade blue steel hands, Keystone silveroid case with a hinged case back, Louis Jacot's "Telephone Dial" crystal guard, factory crown, size 0s, 7 jewels, 3-piece 12mm bund strap.

Value: $1,200–$1,500
Rarity: 10

1916 ELGIN, BOLD ARABIC DIAL

Relumed skeleton hands and factory drilled hour marker lum dots, Illinois solid nickel semi-hermetic case with swivel lugs, Mealy Manufacturing "Duo" crystal guard, factory crown, size 0s, 7 jewels, 2-piece 14mm pigskin strap.

Value: $800–$1,000
Rarity: 9

1917 ELGIN, BOLD ARABIC DIAL

Relumed skeleton hands, Fahys sterling silver "Armored" case with an integrated crystal guard, big size 0s version, replacement crown, 19 jewels, grade 205 transition movement with a solid gold train, 16mm leather strap.

Value: $1,800–$2,000
Rarity: 10 (due to 19 jewel movement)

CHAPTER 10

DIAL DATE STAMPS

Starting in late 1916, the Elgin National Watch Company began date-stamping the dials that they considered to be "military dials." In 1916 and 1917, the datestamps were hand drawn with a pencil or ink on the back side of the military dials. Hand-writing the date stamps continued on the back side of the dials until early 1918 when the pencil or ink was replaced with an actual ink rubber stamp. A date-stamp consisted of a month and a year like "8-18" for August 1918. Elgin continued to date-stamp their "military dials" up through very late 1921 or early 1922. The pencil and ink dates on the back of the dials were very durable and even when the watch got wet ,the date stamp would still be there. The rubber stamp date stamps were a different story, though. If they got wet, there was a very good chance the ink stamp would dissolve completely. If you are taking a watch apart to do a restoration or to service it, do *not* wash the back side of the rubber stamped dials with soap and water or any harsh solvents. This will remove the date stamp and erase the provenance of the dial. Do *not* put the dial in an ultrasonic cleaning machine either or the same will happen. Only clean the front side of the dial. I like to use Dawn dish soap and a Q-tip; this will not destroy the ink date stamp on the back side of the dial. Be sure *not* to wipe the back of the dial when drying either.

Several great examples of Elgin date-stamped dials.

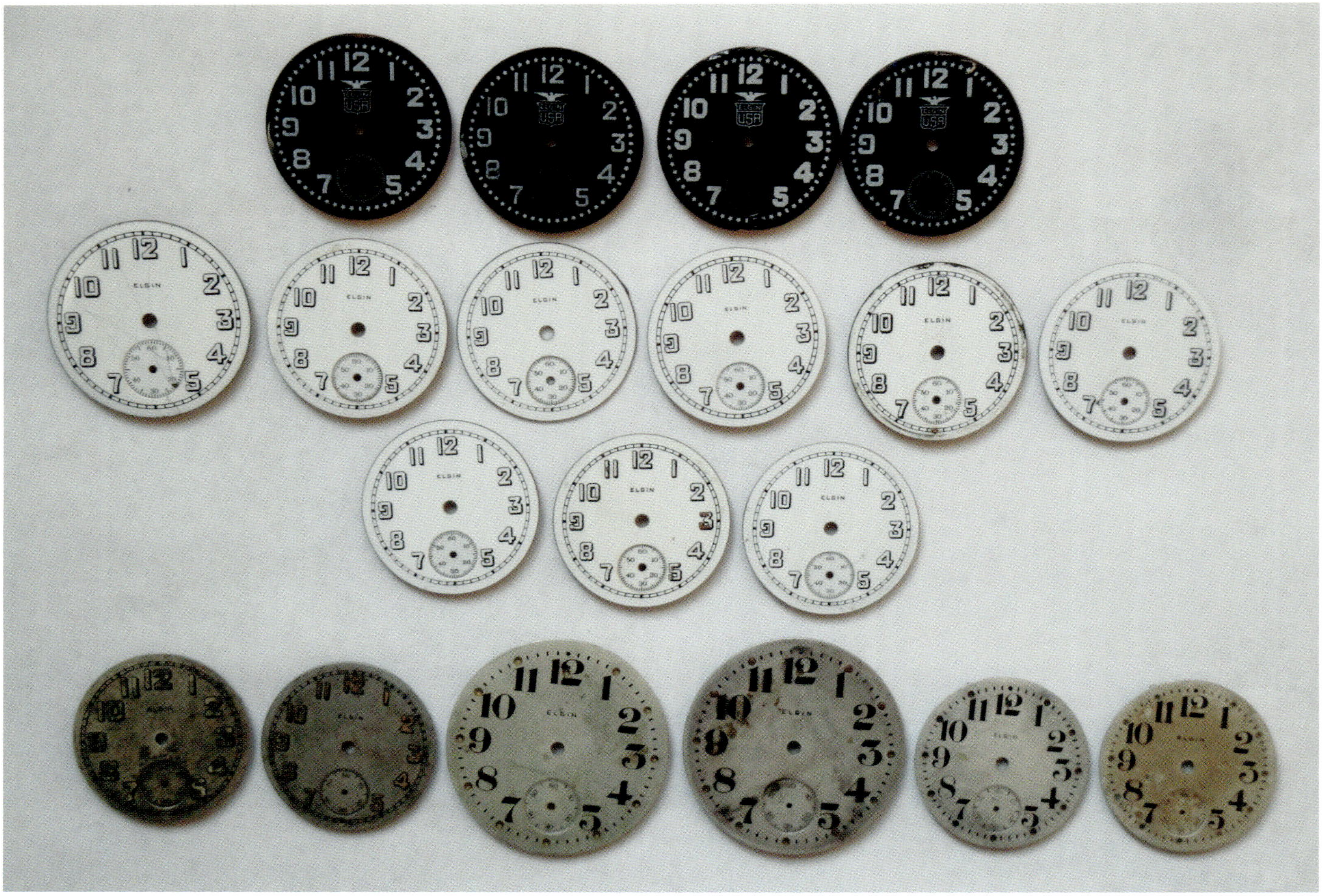

Star Dials, Shadow Box Dials and Bold Arabic Dials

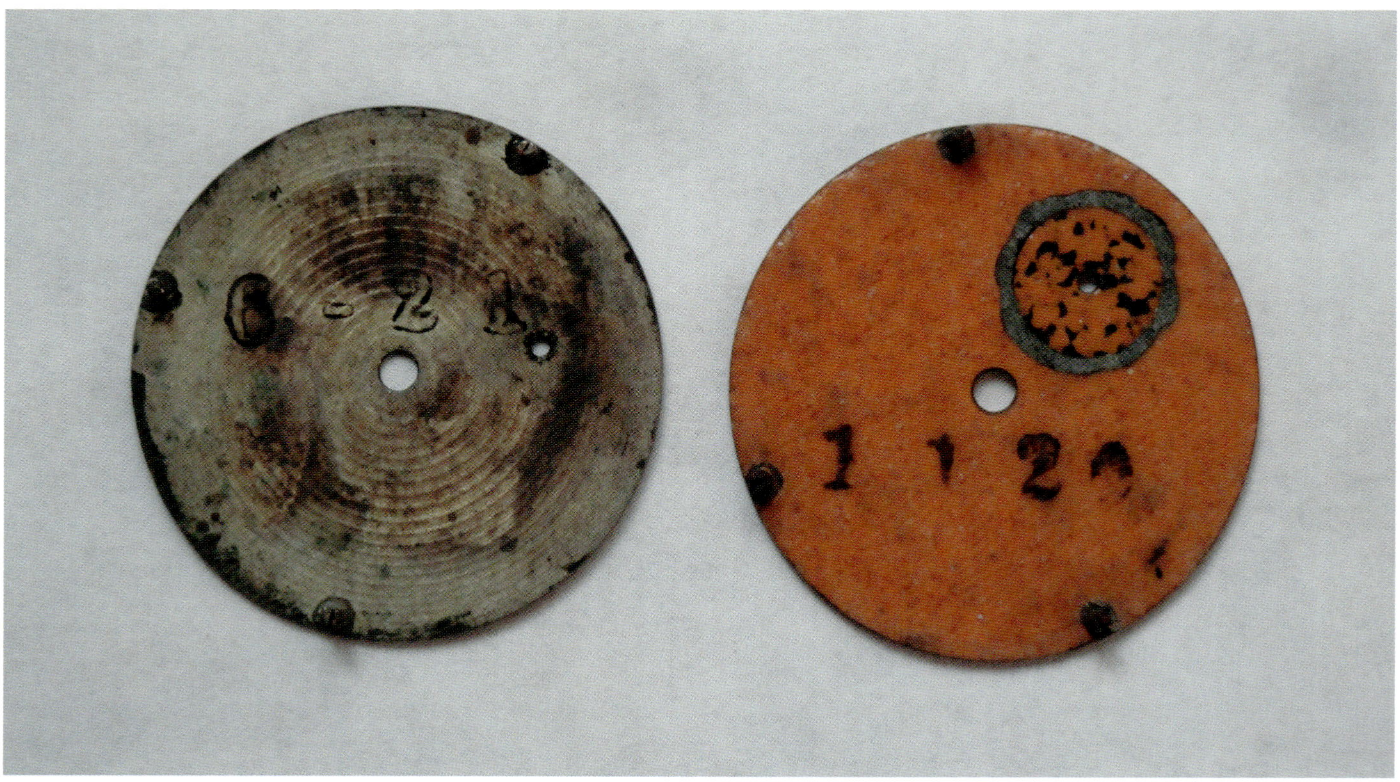

Top row: Rubber-stamped date stamps.
Bottom row: Hand-drawn date stamps.

CHAPTER 11

RECENT RESTORATIONS

Taking a WWI Elgin Trench Watch that is nearly 100 years old that has been neglected for more than 70 years and returning it to splendor is never easy. The images in this chapters show only the results of cosmetic work that makes the watch look fantastic again. What you don't see are the intricate details involved in restoring the watch movement itself, which is more difficult.

Years of experience are needed to know the tolerances of case materials and exactly how far you can push the envelope. Each material—sterling silver, 10K gold filled, solid nickel, etc.—will not turn out the same using the same restoration methods.

The same thing goes for metal dials versus enamel dials: the methods used for restoration differ.

I am not a huge fan of metal dial refinishing. In some cases the dial can be properly restored rather than refinished. Stellar results are not always achieved with every project. Sometimes a watch is just too far gone to be saved, and it then goes into the parts bin for another needy watch. Finding the correct factory parts to finish off a project is quite difficult. Sometimes a project will be shelved for several months or even a year just waiting for the right parts.

Years of research go into learning which hands, dial, crown, strap, and movement go into each watch. Tiny details like these can make or break a watch's value.

This chapter shows what is possible with a watch restoration project; see www.LRFAntiqueWatches.com for closer looks at my step-by-step process.

Before

After

Before

After

Before

After

Before

After

Before

After

Before

After

Before

After

Before

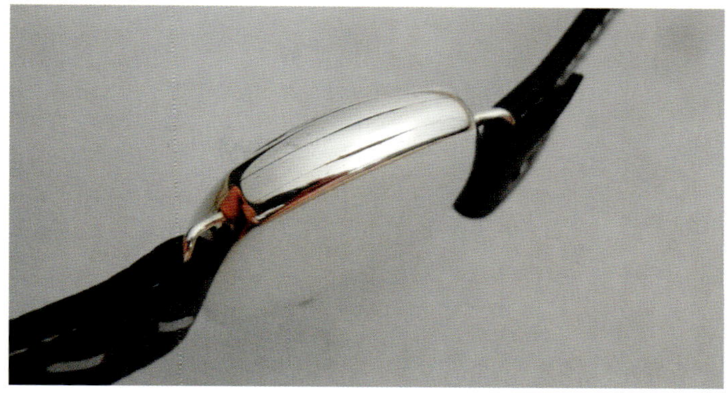

After

Before

After

Before

After

Before

After

CHAPTER 12

ELGIN MOVEMENTS

When I first started collecting Elgin trench watches from World War I, I never opened up the case backs as I was afraid that I might damage the watch somehow. As time passed, and I properly learned how to open the watch cases, I was amazed at what I saw. Another work of art lies underneath the case back! The intricate details that went into each one of these Elgin movements is astounding.

The movements manufactured around the year 1900 were the peak of the demaskeen designs. After that, the movements were still detailed and beautiful, but the amount of detail was decreased each year.

Shown here is one of Elgin's best efforts: 1915, size 3/0s, 17 jewels, raised solid gold jewel chatons, solid gold center wheel and solid gold timing screws. Even the ratchet wheel and crown wheel have very heavy details. Out of the approximately 2,000,000 size 3/0s movements that Elgin manufactured, only 500 of them were made with 17 jewels. This is number 285 of the 500.

Elgin grade 423 movement with 17 jewels; only 500 of these were made.

1899 ELGIN

Size 0s, 19 jewels, jeweled motor barrel, grade 201, solid gold train and raised jewel chatons, serial number 8019412.

1915 ELGIN

Size 3/0s, 15 jewels, grade 419, serial number 18592661.

1917 ELGIN

Size 3/0s, 7 jewels, grade 418, serial number 20616504.

1918 ELGIN

Size 6s, 7 jewels, grade 286, serial number 21100124.

1915 ELGIN

Size 3/0s, 17 jewels, grade 423, solid gold center wheel and raised jewel chatons, serial number 18628285. Elgin only made 500 movements with 17 jewels out the 2,000,000 size 3/0s movements that were produced. Number 285 of 500.

1914 ELGIN

Size 0s, 7 jewels, grade 324, serial number 18294731.

1917 ELGIN

Size 6s, 15 jewels, grade 295, serial number 20000201.

1916 ELGIN

Size 3/0s, 15 jewels, grade 419, serial number 19127337.

1918 ELGIN

Size 3/0s, 7 jewels, grade 463, serial number 21091010.

1898 ELGIN

Size 0s, 19 jewels, grade 205, solid gold train, jeweled motor barrel, serial number 7966999.

1916 ELGIN

Size 3/0s, 7 jewels, grade 442, gilded movement, serial number 19738942.

1916 ELGIN

Size 0s, 15 jewels, grade 354, serial number 19191168.

This was the last year Elgin made size 0s movements.

1913 ELGIN

Size 0s, 7 jewels, grade 320, serial number 17816513.

1917 ELGIN

Size 6s, 7 jewels, grade 286, serial number 20790565.

1917 ELGIN

Size 3/0s 7 jewels, grade 442, gilded movement, serial number 20235405.

1918 ELGIN

Size 3/0s, 7 jewels, grade 464, serial number 21253720.

1917 ELGIN

Size 3/0s, 7 jewels, grade 462, serial number 20154244.

1917 ELGIN

Size 3/0s, 7 jewels, grade 418, serial number 20537894.

1915 ELGIN

Size 3/0s, 7 jewels, grade 413, serial number 18435229.

1911 ELGIN

Size 0s, 17 jewels, grade 264, serial number 16482912.

1914 ELGIN

Size 0s, 7 jewels, grade 320, serial number 18006875.

1914 ELGIN

Size 0s, 15 jewels, grade 354, serial number 18306979.

1917 ELGIN

Size 3/0s, 7 jewels, grade 418, serial number 20538130.

1918 ELGIN

Size 3/0s, 15 jewels, grade 464, serial number 21313209.

CHAPTER 13

STRAPS, BUCKLES & FASTENERS

Some very interesting aspects of Great War watches are their original straps, buckles, and fasteners. A wide range of products were available to fasten watches to the wrist. Olive drab webbing, pigskin leather, and cowhide leather were the most popular straps. Buckles and fasteners were made from a variety of metals, such as nickel, sterling silver, gold filled, solid gold, nickel silver, and gun metal.

Khaki webbing had several grades of thread count, thickness, and overall quality. I have been able to inspect dozens of khaki straps that have survived into modern times, and in my opinion, the manufacturer that stood above the rest was the J.F. Sturdy's Sons Company in Attleboro, Massachusetts. Their strap and fastener designs were of the finest quality and command top dollar in today's market.

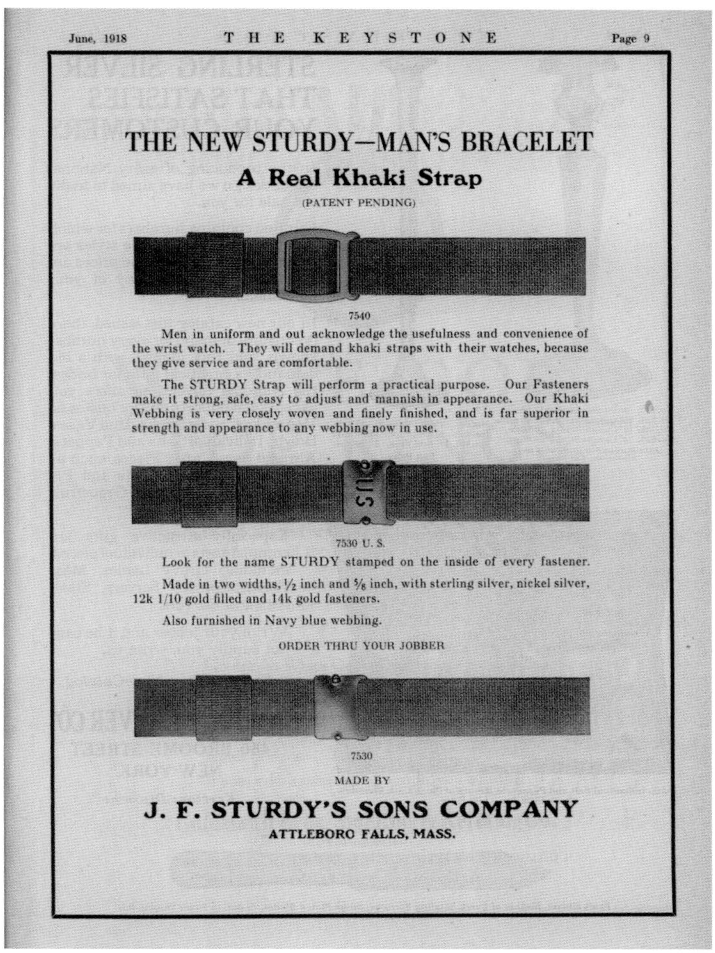
June, 1918 THE KEYSTONE Page 9

THE NEW STURDY—MAN'S BRACELET

A Real Khaki Strap

(PATENT PENDING)

7540

Men in uniform and out acknowledge the usefulness and convenience of the wrist watch. They will demand khaki straps with their watches, because they give service and are comfortable.

The STURDY Strap will perform a practical purpose. Our Fasteners make it strong, safe, easy to adjust and mannish in appearance. Our Khaki Webbing is very closely woven and finely finished, and is far superior in strength and appearance to any webbing now in use.

7530 U. S.

Look for the name STURDY stamped on the inside of every fastener.

Made in two widths, ½ inch and ⅝ inch, with sterling silver, nickel silver, 12k 1/10 gold filled and 14k gold fasteners.

Also furnished in Navy blue webbing.

ORDER THRU YOUR JOBBER

7530

MADE BY

J. F. STURDY'S SONS COMPANY

ATTLEBORO FALLS, MASS.

The J.F. Sturdy's Sons Company spared no expense when it came to their marketing campaigns. Their advertisements were precise and to the point, with a patriotic theme.

Of the J.F. Sturdy's fasteners, the "US" stamped version is the most desirable. One nickel version recently sold at auction for $350; this was just for the original khaki strap and fastener, no watch included. Only one 14K solid gold "US" fastener has ever been documented, and it is shown below. I would put its value at $1,000–$1,500.

An advertisement for J.F. Sturdy's US stamped straps.

This is the only known J.F. Sturdy's 14K solid gold "US" fastener that exists.

J.F. Sturdy's also had different colors of khaki webbing for different services. Khaki Green for the Army and Blue for the Navy.

With eminent victory looming for the Allies only six weeks away J.F. Sturdy's decided to take a poke at the Kaiser in this advertisement. Note that the American soldier is wearing a wristwatch and the Kaiser is not.

Khaki straps had two basic designs. One was a looped strap.

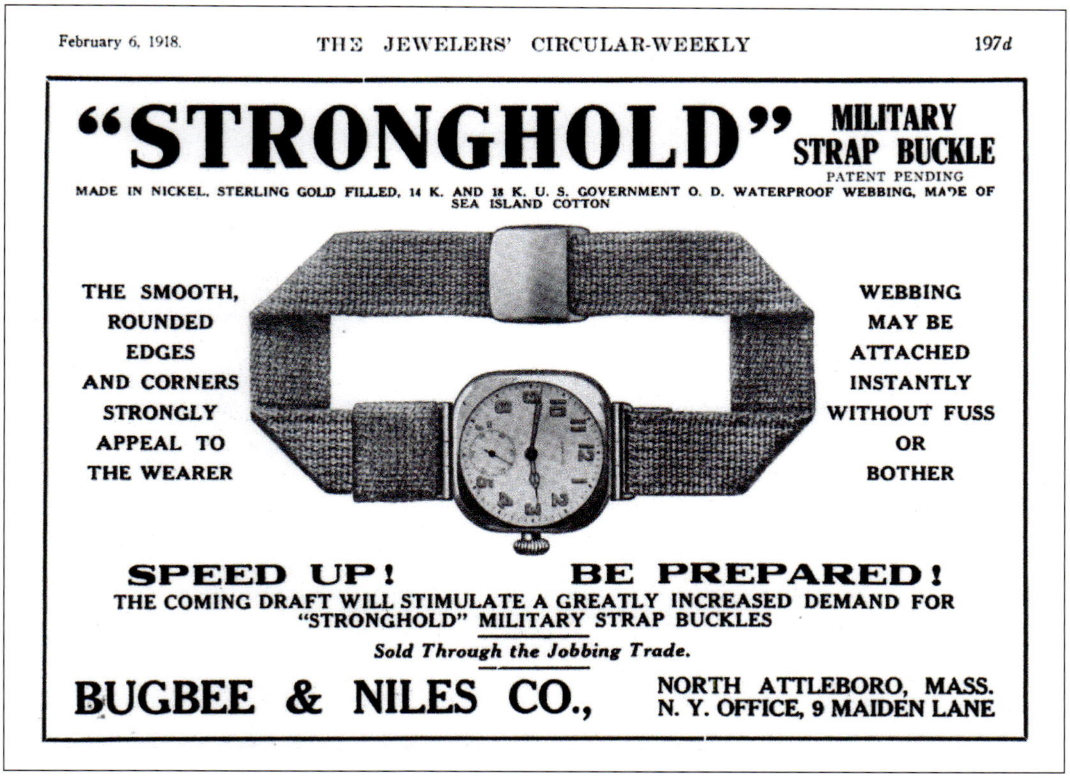

This strap was looped through the top or bottom lug. Note the thicker end of the strap near the 6 o'clock position. The entire strap would be looped through this portion.

Here is an example of an original Bugbee & Niles strap buckle that was date-stamped July 16, 1918. This is a two- piece, snap-together "Military Strap Buckle."

The other strap design was simpler and straight. The strap went underneath the case back directly through the two solid lugs. Sussfeld, Lorsch & Company was a smaller firm, but they had the name "Pershing" attached to their product.

The straight strap design.

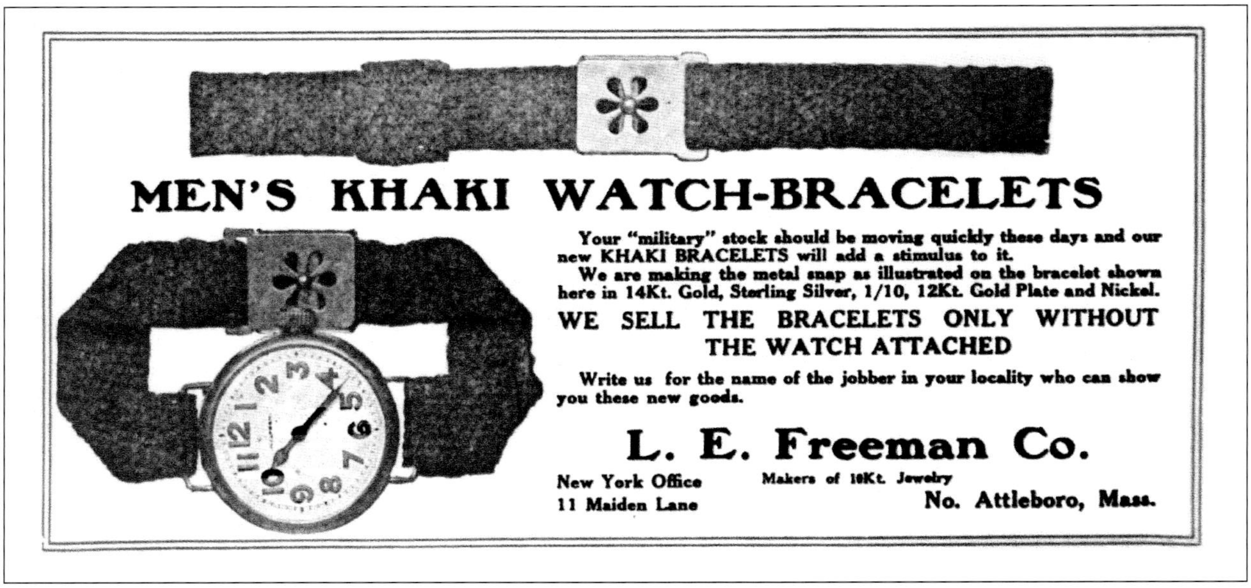

This is another example of a "looped" design manufactured by the L.E. Freeman Company of North Attleboro, Massachusetts. This general area in Massachusetts seemed to be the hub for strap and fastener manufacturing, as several different companies were located here.

88 THE JEWELERS' CIRCULAR-WEEKLY April 3, 1918.

The new

Liberty Military Strap

(PATENTS APPLIED FOR)

FOR WRIST WATCHES

The most popular wrist watch strap ever manufactured—the quickest seller. Made of the best grade of stout webbing, insuring lengthy and satisfactory service. Easily attached to any wrist watch in a short time.

Especially adapted for the boys "over there."

SIMPLE—CONVENIENT—DURABLE

The slightest pull adjusts it firmly on the wrist as tightly as desired

KHAKI COLOR WEBBING

Sterling Silver Slide

Place your order now so as to obtain prompt service and an early shipment. Your stock will not be complete and up-to-date without them. Sell a few extra with each military watch.

MADE IN 7/16 AND 5/8 INCH WIDTHS

If your jobber has not yet received his supply, we will send them to you direct.

$4.00 DISPLAY BOX OF ONE DOZEN

MANUFACTURED SOLELY BY
WATCH SPECIALTIES CO.
DEPT. OF **GRUEN WATCHMAKERS' GUILD**
Time Hill
Cincinnati, Ohio - - - - U. S. A.

You don't take it OFF— you slide it *UP*.

Watch Specialties Company, a division of the Gruen Watchmakers' Guild, also manufactured a strap called the "Liberty Military Strap." It featured a "sterling silver slide"; the wearer simply pulled on the slide to tighten the watch on the wrist. Note the stamped image of a wristwatch on the slide.

The Max Hirsch Company specialized in a variety of leather straps. Several different versions of the "Kitchener: strap were available in a variety of lengths to accommodate the infantry soldier or the aviator pilot. Aviators would require a longer strap as the watch would normally be worn on the outside of a thick jacket and gloves. The infantry solder would require a shorter strap as he would wear his wristwatch directly on the skin in most cases.

This May of 1918 advertisement from the Max Hirsch Company depicts a bi-plane and an aviator in full uniform sporting a "Kitchener" strap.

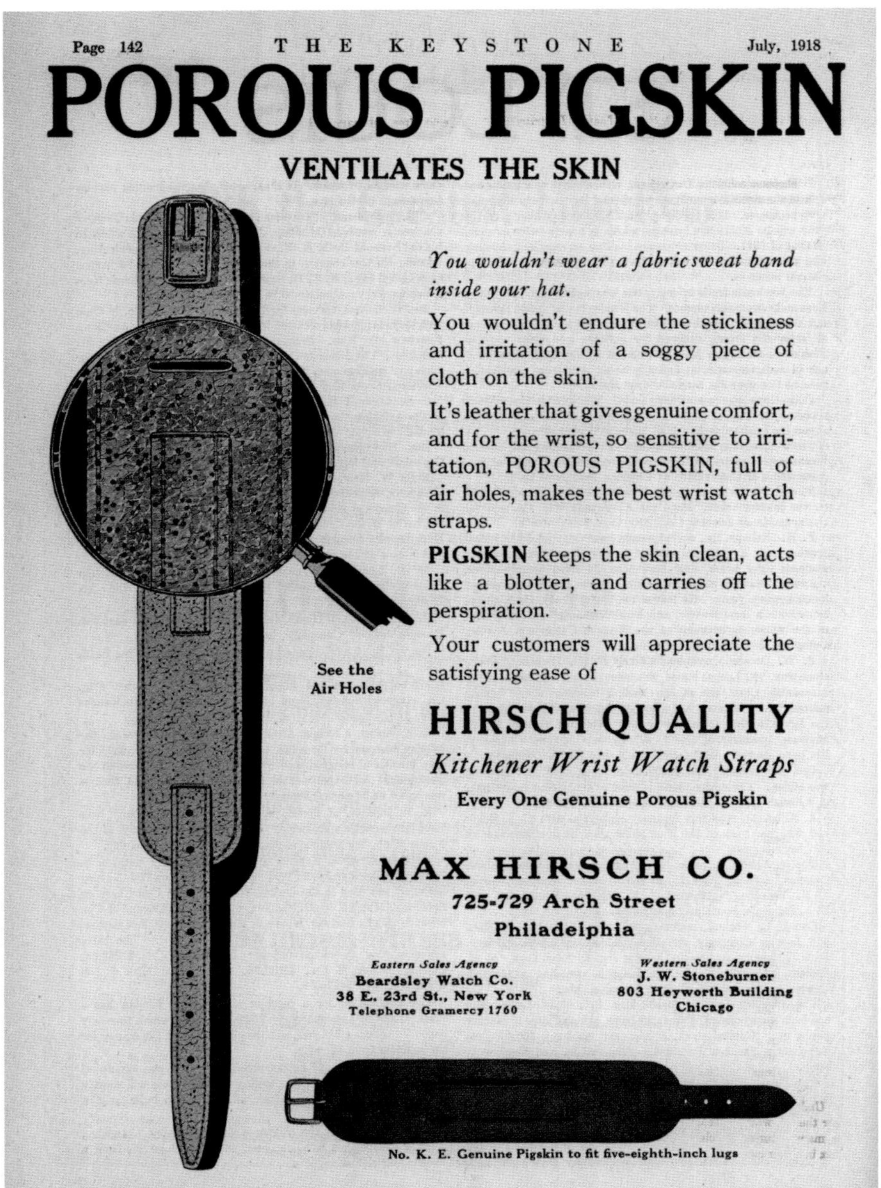

A July 1918 advertisement from Hirsch focuses on the porous pigskin leather "Kitchener" strap, ventilated with air holes for the wearer's comfort.

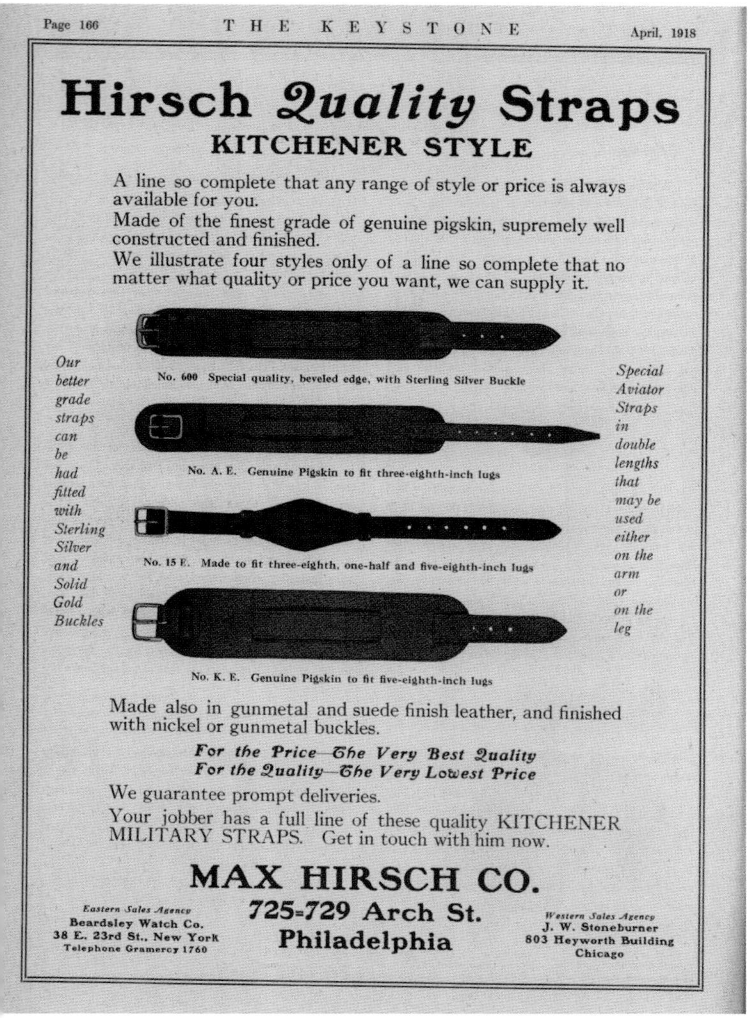

This April1918 advertisement from Hirsch shows four different styles of the "Kitchener" strap in pigskin and suede leather.

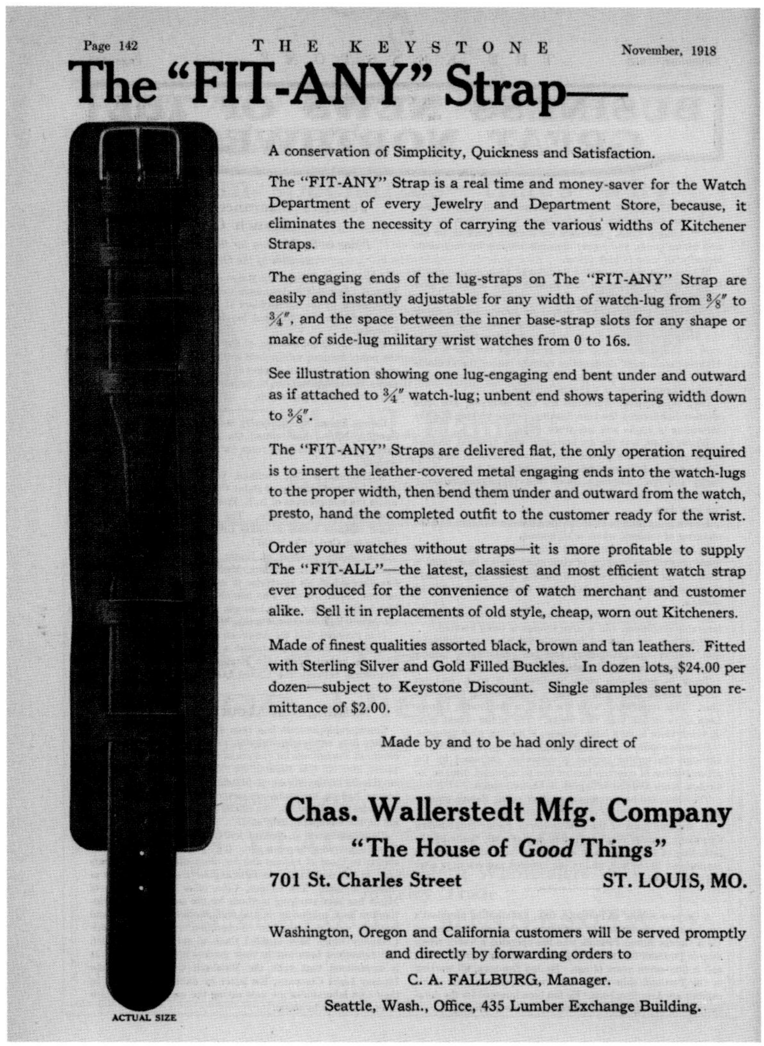

Chas. Wallerstedt Manufacturing Company of St. Louis, Missouri, made a "FIT-ANY" three-piece strap; the Hirsch straps had just two pieces.

The Illinois Watch Case Company of Elgin, Illinois, was mostly known for their high-quality watch cases, but they also made leather straps. Adding a compass to the strap was very practical for the soldier in the field, and being able to have a picture of a loved one back home would be quite comforting.

At right are two leather Kitchener style straps with the original buckles from the Great War era. These are historically important but very old and extremely brittle. The stitching around the buckles has become very weak and is starting to disintegrate. The odor from these straps is not very pleasant either. Note that the strap on the left is one inch longer than the one on the right; it was probably made to be worn over clothing instead of directly on the wrist.

Page 144 THE KEYSTONE October, 1918

Our Latest Creations in Wrist Watches for Uncle Sam's Boys

Solid Gold—Gold Filled
Sterling Silver

New Added Feature

COMPASS or PHOTOGRAPH EQUIPMENT
on Strap, Supplied with any Model

ROUND—SQUARE—OCTAGON and
CUSHION SHAPES

Protected Hunting Military Wrist Watch Case

Can be readily operated with a single motion of one hand
THE MOST SECURE MOVEMENT PROTECTOR

12s—Os—3/Os—and Jewel Series to Fit
Elgin, Waltham, Illinois and all Standard American and Swiss Movements

ASK YOUR JOBBER TO SHOW YOU OUR
GENERAL LINE OF
ALL KINDS OF WATCH CASES

ILLINOIS WATCH CASE CO.

Factory at ELGIN, ILLINOIS

15 Maiden Lane NEW YORK | 704 Market Street SAN FRANCISCO | 904 Heyworth Bldg. CHICAGO

Today, these straps command top dollar; I have seen them sell for $300-$400 without the watch included.

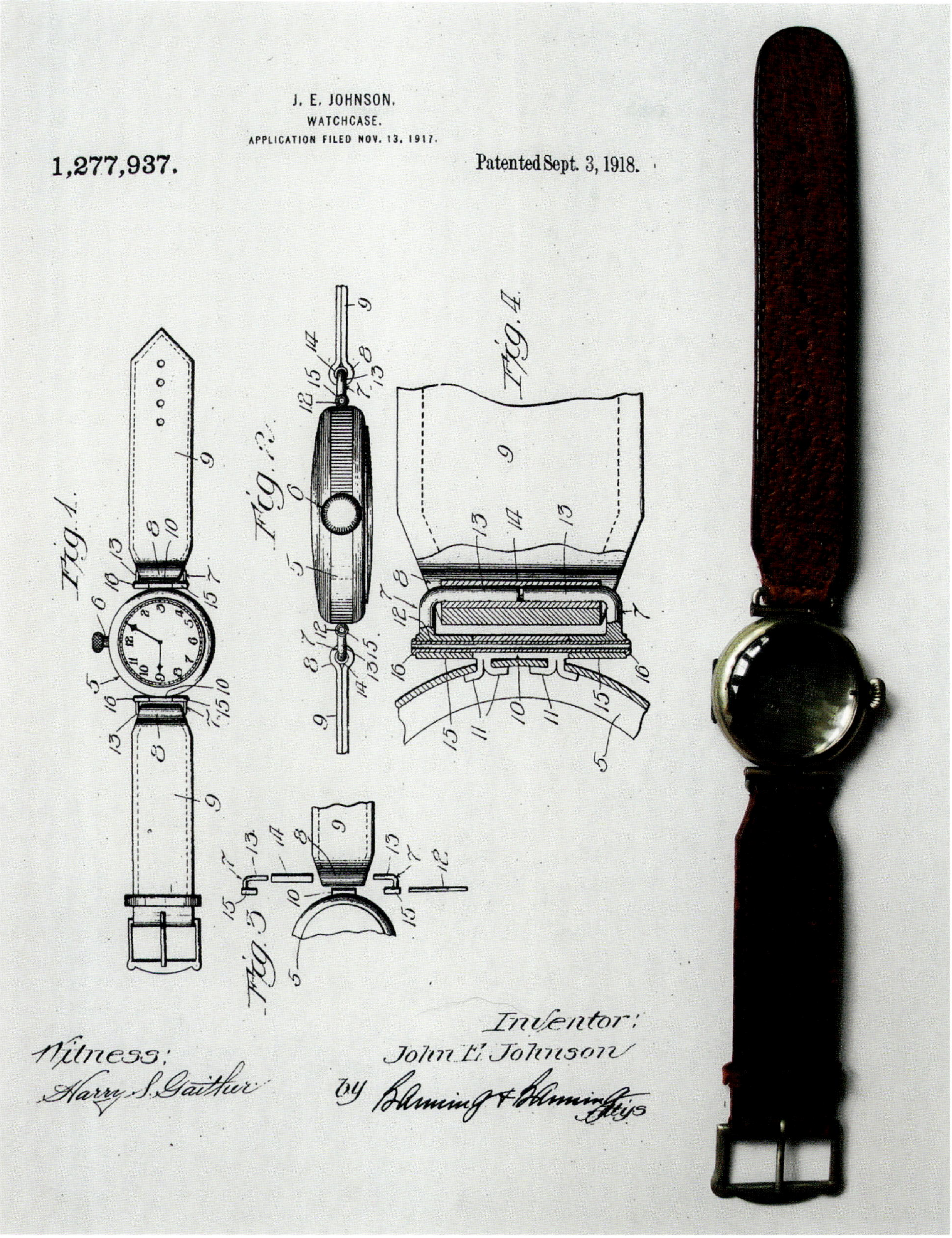

A telltale sign of an original strap that was made for cases from the Illinois Watch Case Company is the tapered design. The strap will be wide at the buckle end but will then taper down near the lugs. This example has the original factory buckle as well. It measures 20mm at the buckle end then tapers to 14mm at the swivel lugs. It is pretty rare to find these straps still intact as the leather disintegrates with age. This is a show piece only; the leather is too brittle to be worn.

CHAPTER 14

ODDS & ENDS

This chapter explores various tidbits about Elgin trench watches and their dials, cases, and more.

From civilian to military dial

Have you ever seen a paper dial overlay that converts a regular civilian dial into a military type dial with radium lum?

Black Star Dial fakes

Did Elgin manufacture the Black Star Dial in metal? No, they did not, and if you ever see one like this, it's a fake or has been refinished.

Look at the "V" notch just above the 12; it's a telltale sign of refinishing.

I have also seen fake Black Star Dials made from plastic. They have a very glossy finish and are kind of hard to spot, but one telltale sign is that the dial foot at the 38 minute position is off slightly.

At left is a real Elgin Black Star Dial; they were only made in enamel.

This is the most perfect example I have ever seen. It was stuck from a brand new die back in 1918. Look at the exquisite detail on the eagle's wings.

A night and day difference from the poorly refinished dial on the fake.

Jumping spacers

Since Elgin stopped manufacturing size 0s movements in late 1916, the need for a "size-jumping spacer" was created. With a size-jumping spacer, you could put any size 3/0s Elgin movement into an American size 0s watch case. Elgin was still manufacturing size 0s military dials until 1918.

A common size 0s dust ring is on the left of each photograph at left, and a size 3/0s to a size 0s size-jumping spacer on the right. Notice the large lip on top of the size-jumping spacer. The common dust ring on the left is much thinner and does not have a dial lip.

Nickel trench watch cases

Ever wonder who made the "no-name" nickel trench watch cases without a maker's mark? It was the Wadsworth Watch Case Company of Dayton, Kentucky. Wadsworth was known as a high- end case maker specializing in solid gold and gold-filled cases. They did mark their sterling silver trench watch cases, but for some odd reason they did not mark the solid nickel cases. They probably did not want their name on a basic nickel watch case. I have seen only one of these cases from Wadsworth that was solid nickel with their name stamped on the inside of the case back; all of the others are just stamped "nickel" with a serial number. These nickel cases are very well made!

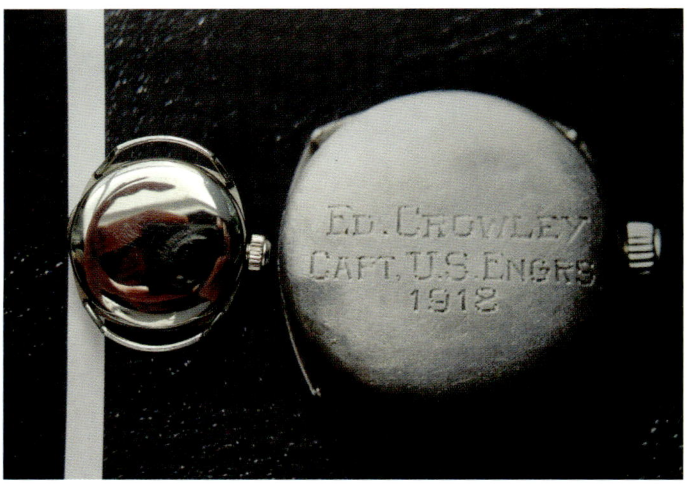

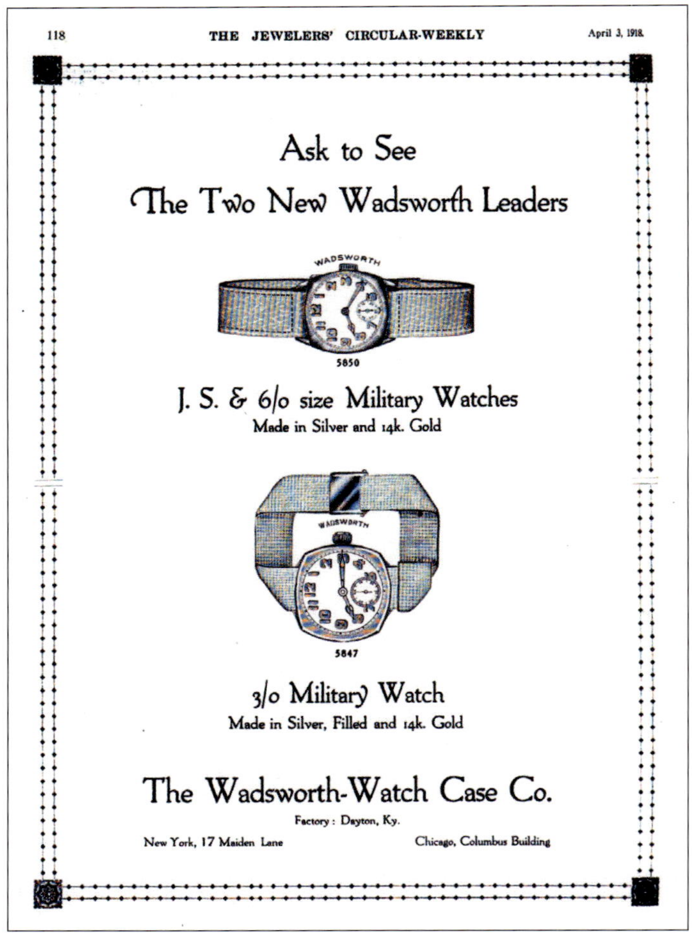

118 THE JEWELERS' CIRCULAR-WEEKLY April 3, 1918.

Ask to See

The Two New Wadsworth Leaders

WADSWORTH

5850

J. S. & 6/0 size Military Watches

Made in Silver and 14k. Gold

WADSWORTH

5847

3/0 Military Watch

Made in Silver, Filled and 14k. Gold

The Wadsworth-Watch Case Co.

Factory: Dayton, Ky.

New York, 17 Maiden Lane — Chicago, Columbus Building

Top left: Note that the lugs are attached to the bezel and not to the center of the case.

Post-war watch sales

The photograph at right is probably my all-time favorite from the Great War era. This picture was taken in very early 1919 in Washington, D.C. After the Armistice was signed in November 1918, military wristwatches were sold off at a fraction of their original value. A $15 gold-filled trench watch was sold for only $4 and leather "Kitchener" straps could be had for only 10 cents. Copies of this photograph are available at the National Archives in Washington, D.C.

Courtesy Library of Congress.

A rare 12 o'clock hinged cover

There are probably many American watch cases that have yet to be documented. It's possible that some cases were prototypes, or they possibly had a very limited production runs. This case made by the Illinois Watch Case Company is a prime example. When brand new, this sterling silver case had a hinged cover at the 12 o'clock position, which has been lost in time. It is unknown what this cover looked like, and we can only speculate. The Illinois Watch Case Company made others cases similar to this with a solid cover, but the hinge was at the 9 o'clock position rather than at the 12 o'clock position. The hinged cover on this case could have been solid, pierced, or in the demi-hunter style perhaps. Below is an October 1918 advertisement from the IWC Co that shows a similar watch case with a solid cover, but the cover hinge is in the 9 o'clock position. The Swiss made similar cases with hinged covers, but I have not seen one fully intact made by an American case manufacturer.

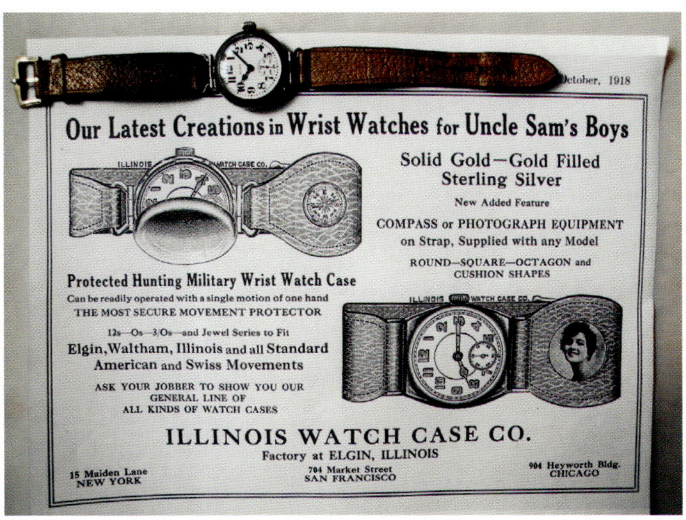

October, 1918

Our Latest Creations in Wrist Watches for Uncle Sam's Boys

Solid Gold—Gold Filled
Sterling Silver

New Added Feature

COMPASS or PHOTOGRAPH EQUIPMENT
on Strap, Supplied with any Model

ROUND—SQUARE—OCTAGON and
CUSHION SHAPES

Protected Hunting Military Wrist Watch Case

Can be readily operated with a single motion of one hand
THE MOST SECURE MOVEMENT PROTECTOR

12s—Os—3/Os—and Jewel Series to Fit
Elgin, Waltham, Illinois and all Standard
American and Swiss Movements

ASK YOUR JOBBER TO SHOW YOU OUR
GENERAL LINE OF
ALL KINDS OF WATCH CASES

ILLINOIS WATCH CASE CO.

Factory at ELGIN, ILLINOIS

15 Maiden Lane
NEW YORK

704 Market Street
SAN FRANCISCO

904 Heyworth Bldg.
CHICAGO

These images show the 12 o'clock hinge position. This 1918 Elgin would be quite valuable if fully intact; I would estimate its value at around $2,000 plus if the cover were present. This is the original IWC Co factory band.

The Big Size 12s

It is a common misconception that the largest American trench watches made during the Great War were the giant size 6s. The largest were, in fact, the behemoth size 12s. At least two American case manufacturers made trench watch cases for size 12s movements: The Philadelphia Watch Case Company and the Illinois Watch Case Company. The Philadelphia Watch Case Company did very little advertising for their wristwatch cases during this period, so the hope of finding a 12s wristwatch case ad are very slim. But the Illinois Watch Case Company had at least four advertisements for their size 12s wristwatch cases, and they are shown here. All four are from 1918. There was a shortage of size 3/0s movements in 1918 due to the war effort, so the case manufacturers probably began making 12s wristwatch cases in order to keep up with demand; size 12s movements were still plentiful.

The size 12s American trench watches are the rarest of the rare. I have only seen two of them. One was an Elgin in a square Illinois gold-filled case, but at the time I did not think that is was genuine. I thought it was a modified pocket watch case, but when I saw the advertisement below at a later date I knew that I had made a mistake by taking a pass on the watch, as the lugs were correct. The size 12s model that I saw is illustrated at the bottom right of the advertisement shown below at top left. This case measures approximately 50mm lug to lug and 44mm without the crown. It is massive.

It is unknown why the Illinois Watch Case Company put so much time and effort into the size 12s wristwatch case advertisements when so few were actually made.

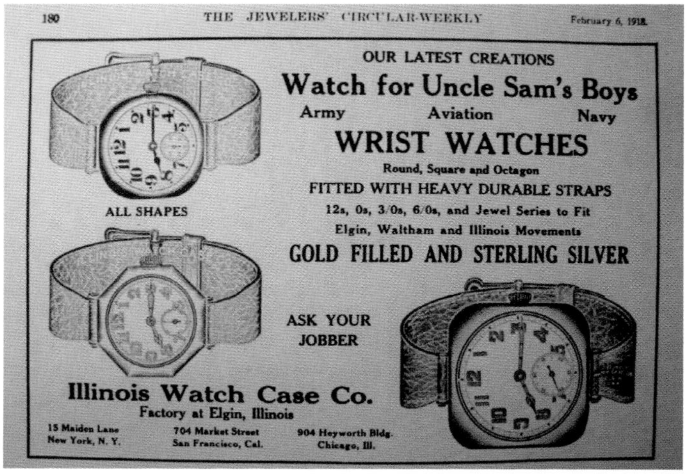

The second size 12s American trench watch that I have seen was made by the Crown Watch Company. Crown Watch was owned by the Keystone Watch Case Company, which also owned the Philadelphia Watch Case Company and Bates & Bacon (B&B). This Crown movement is housed in a Philadelphia semi-hermetic gold- filled case. The watch is physical proof that the behemoth size 12s American trench watch does in fact exist. The image at right shows four different sizes of the exact same case design under the Keystone Watch Case Company family of companies; only the case materials are different.

All four cases shown here were available in semi-hermetic, which is a threaded case, or in the more common snap-together cases. The winding crown style was a variable depending on model year. Look closely at the lug design on all four watches. They all have the exact same 16mm lug design, unmistakably from the Keystone family of companies.

I estimate the value of a size 12 Elgin Trench Watch at $3,000–$5,000

Rarity: 10+

That estimate would be for a gold-filled case and with the correct enamel military dial and hands. The lugs would have to factory installed. Beware of converted pocket watch cases that have had the lugs added after the fact by a jeweler. Shady dealers will try to pawn these off as genuine when they are basically worthless. Have the watch inspected by an expert first.

From left to right:
Size 3/0s Elgin in a Philadelphia silverode case. 36mm x 31.4mm x 16mm
Size 0s Elgin in a Philadelphia silverode case. 37mm x 32.6mm x 16mm
Size 6s Elgin in a Bates & Bacon 10K gold filled case. 42mm x 38mm x 16mm
Size 12s Crown in a Philadelphia 10K gold filled case. 53mm x 47mm x 16mm

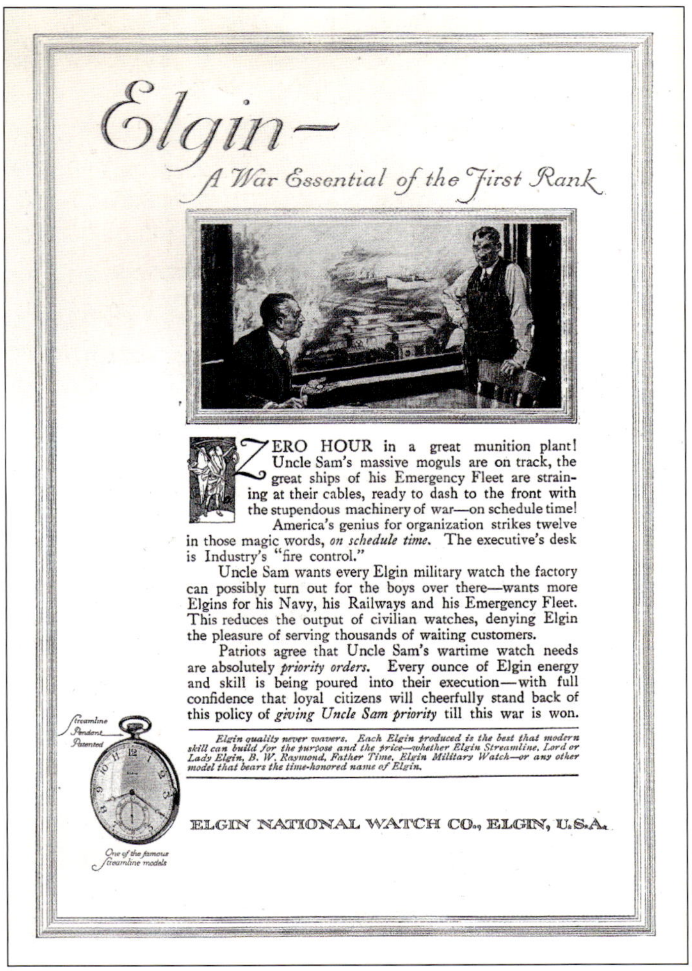

A wartime rally

I found th Elgin advertisement at left in a bin along with several others a few years ago. Judging from the content and the design of the pocket watch that is illustrated, I would date it back to mid-1918. When the United States entered the Great War in 1917, the vast majority of available Elgin military wristwatches were bought by the U.S. Army. They could also be bought at local shops or through mail order, but supplies were low at times. The same thing happened again during WWII: the needs of the general public were put on the back burner so production could be dedicated to the military.

Proper names for watches

On page 6, I discussed "proper names" being given to Elgin wristwatches from the Elgin National Watch Company in February 1919. The very first "named" Elgin wristwatches from the factory were the "Pershing," "Foch," and "Kitchener." But, this document gives proper names to watch cases from the Illinois Watch Case Company dating back to September 5, 1917.

This document is very special as it is the only advertisement from the IWC Co that I have found that lists the proper names of these cases. I consulted with several watch experts from this era, and we agreed that it would be correct to call these watches by the proper names given to them by the Illinois Watch Case Company, even though they had nothing to do with manufacturing the movements. Usually the watch movement factory gives a proper name to a particular wristwatch, but in this case we have an exception.

For example, it would be correct to refer to an Elgin with an octagonal case as an Elgin "Admiral Evans," and if it had an Illinois movement, it would be called an "Illinois Admiral Evans." It would be a "Waltham Admiral Evans" if it featured a Waltham movement. The same can be said for the "Admiral Benson," "General Funston," and the "General Joffre."

An ad in the Jewelers' Circular of September 5, 1917, gives proper names to some watch cases.

Original factory box of cases

Shown below is an original factory box full of watch cases from the Illinois Watch Case Company. This is how the watch cases were packaged and shipped to the watchmakers and jobbers during the Great War era. Each box of Illinois cases contained six watch cases and each watch case was stored in a cotton pouch. Each box had six cardboard compartments for the cases. These boxes were very well made and pretty strong. The boxes were not date-stamped but judging from the case style they are from 1917. This particular box holds semi-hermetic cases, and all six of them have never been used: they have never had a movement installed in them. The top of the box features a picture of the Illinois Watch Case Company factory building. One side of the box has text that describes what type of cases the box holds: solid nickel cases, size 0s, bracelet-style case with wire loop lugs, open face. S.B.B. stands for screw-bezel-back, as these are semi-hermetic threaded cases so the bezel and case back screw on and off rather than popping on and off. I have only seen one other complete box of unused Illinois cases in my career. Finding a complete box of virgin cases fully intact is quite rare. It is amazing that this case has survived about ninety-seven years without ever being used.

CHAPTER 15

CASE GRADES

Solid gold and sterling silver case grades will clearly be marked for content. The following lists are for gold-filled, gold-plated and alloy case grades.

American Watch Case Company

Cashier: Gold-filled case guaranteed for 25 years
Fortune: Gold-filled case
Empress: Gold-filled case
Silveroid: Nickel alloy
Derby: Rolled gold plate
Nickel Silver: Nickel alloy

Bates & Bacon

Favorite: Gold-filled case guaranteed for 25 years
Puritan: Gold-filled case guaranteed for 5 years
Regal: Gold-filled case guaranteed for 10 years
Royal: Gold-filled case guaranteed for 20 years
Peer: Gold-filled case guaranteed for 20 years
Gem: Gold-filled case

Brooklyn Watch Case Company

Bristol: 14K gold-filled case, guaranteed for 25 years
Windsor: Gold-filled case, guaranteed for 20 years

Crescent Watch Case Company

Extra: 14K gold-filled case
Star & Crescent Moon Trade Mark: Gold-filled case, guaranteed for 20 years
Planet: Gold-filled case, guaranteed for 30 years

Dennison Watch Case Company

Sun Trade Mark: Gold-filled, guaranteed for 25 years
Moon Trade Mark: Gold-filled, guaranteed for 20 years
Star Trade Mark: Gold-filled, guaranteed for 10 years
Nickel: Solid nickel

Dueber Watch Case Manufacturing Company

Special, Anchor-in-Shield: 14K gold-filled, guaranteed for 25 years
Champion: 14K gold-filled, guaranteed for 20 years
Anchor and Serpent: 14K gold-filled
Gladiator: 10K gold-filled, guaranteed for 20 years if so marked, 10 years if not marked
Anchor and Serpent, Newport: 10K gold-filled
Silverine: Nickel alloy

Joseph Fahys & Company

Warranted To Wear Permanently: 14K gold-filled, no time limit on warranty
Bristol: 14K gold-filled, guaranteed for 25 years
Monarch: Gold-filled, guaranteed for 20 years
Conqueror: 12K gold-filled, guaranteed for 20 years
Montauk: Gold-filled, guaranteed for 15 years
Oresilver: Nickel alloy case

Illinois Watch Case Company

Arrow: Base metal, chrome plate, golden color, 1920s
Bona Fide: Gold-filled, guaranteed for 20 years
Commander: Gold-filled, guaranteed for 20 years
Elgin Giant: 10K gold-filled, guaranteed for 20
Elgin Pride: 16K gold-filled, guaranteed for 25 years (25 percent gold weight)
Elgin Tiger: Rolled gold plate, guaranteed for 5 Years
Empire: Gold-gilled, guaranteed for 25 years.
Famous: Gold-colored base metal

Monitor: Gold-filled, guaranteed for 5 years
Napoleon: Gold-filled, guaranteed for 25 years
Nickel Silver: Nickel alloy
Nubian: Gun metal
Pride-of-Elgin: 16K gold-filled, guaranteed for 25 years (25 percent gold weight)
Ramona: Gold-filled, double stock, guaranteed for 20 years
Royal: Gold tone base metal
Spartan: Base metal, chrome-plate, silver color
Supreme: Gold-filled, guaranteed for 10 years
Supreme: Base metal, chrome-plate, 1920s
Tivoli: 14K gold-filled, post-1924
Winner, Horse Head Trade Mark: Gold-filled, guaranteed for 20 years

Keystone Watch Case Company

Keystone Extra: Gold-filled, higher grade (thicker gold) than J. Boss
J. Boss, scale & crown trade mark: Gold-filled, guaranteed for 25 years

J. Boss, plain balance scale trade mark: Gold-filled, guaranteed for 20 years
J. Boss, in oval Keystone trade mark: 10K gold-filled, guaranteed for 15 years
Keystone name in arch over Keystone trade mark: Gold-filled, guaranteed for 15 years
Keystone name in arch over Star & Keystone trade mark: 10K gold-gilled
Victory: Rolled gold plate, guaranteed for 10 years
Cyclone: Rolled gold plate, guaranteed for 5 years
Bee Hive: Rolled gold plate, guaranteed for 5 years
Nickeloid: Nickel alloy

Philadelphia Watch Case Co

Crown: Gold-filled, guaranteed for 25 years
Lion: Gold-filled, guaranteed for 20 years
Royal: Gold-filled, guaranteed for 20 years
Victory: Gold-filled, guaranteed for 10 years
Premier: Gold-filled, guaranteed for 5 years
President: Gold-filled, guaranteed for 3 years
Paramount: Rolled gold, guaranteed for 1 year
Perfection: 14K gold plated
Silverode: Nickel alloy

Star Watch Case Company

Star: Gold-filled, guaranteed for 25 years
Defiance: Gold-filled, guaranteed for 10 years
Majestic: Gold-filled, guaranteed for 5 years
Stellar: 10K rolled gold plate case
Scepter: Gold-filled case
Starwhite: Gold-filled case

Wadsworth Watch Case Company

Permanent: Gold-filled, guaranteed permanently
Pilot: Gold-filled, guaranteed for 25 years
Special: Gold-filled, warranted for 20 years
Referee: Gold-filled, guaranteed for 20 years

CHAPTER 16

CUMULATVIE PRODUCTION NUMBERS

These are the estimated cumulative production numbers for Elgin watch movements. The dates listed are only an estimate, as some serial numbers were assigned but they weren't produced in the exact numerical order. For example, an Elgin watch movement with a serial number that starts with a "19" was approximately produced in the year 1916, while a serial number that starts with the number "20" was approximately produced in 1917, and a serial number that starts with the number "21" was produced around 1918. These estimates were taken from elginwatches.org. This is the most accurate date list I have seen.

1867	101
1868	31,000
1869	71,000
1870	101,000
1871	126,000
1872	152,000
1873	176,000
1874	210,000
1875	310,000
1876	410,000
1877	510,000
1878	552,000
1879	601,000
1880	701,000
1881	801,000
1882	1,000,000
1883	1,440,000
1884	1,650,000
1885	1,850,000
1886	2,000,000
1887	2,550,000
1888	3,000,000
1889	3,550,000

1890	4,000,000
1891	4,400,000
1892	4,890,000
1893	5,000,000
1894	5,500,000
1895	6,000,000
1896	6,550,000
1897	7,000,000
1898	7,550,000
1899	8,100,000
1900	9,100,000
1901	9,350,000
1902	9,755,000
1903	10,100,000
1904	11,000,000
1905	12,100,000
1906	12,500,000
1907	13,100,000
1908	13,550,000
1909	14,100,000
1910	15,100,000
1911	16,000,000
1912	17,000,000

1913	17,550,000
1914	18,000,000
1915	18,500,000
1916	19,000,000
1917	20,000,000
1918	21,000,000
1919	22,000,000
1920	23,000,000
1921	24,000,000
1922	25,000,000
1923	26,000,000
1924	27,000,000
1925	28,000,000
1926	29,000,000
1927	30,000,000
1928	32,000,000
1929	33,000,000
1930	33,300,000
1931	33,500,000
1932	33,700,000
1933	34,000,000
1934	35,000,000
1935	35,500,000

1936	36,200,000
1937	37,000,000
1938	37,900,000
1939	38,200,000
1940	39,100,000
1941	40,200,000
1942	41,100,000
1943	42,200,000
1944	42,600,000
1945	43,200,000
1946	44,000,000
1947	45,000,000
1948	46,000,000
1949	47,000,000
1950	48,000,000
1951	50,000,000
1952	52,000,000
1953	53,300,000
1954	54,000,000
1955	54,500,000
1956	55,000,000